New Perspectives in German Studies

General Editors: Professor Michael Butler, Head of the Department of German Studies, University of Birmingham and Professor William Paterson, Director of the Institute of German Studies, University of Birmingham

Over the last twenty years the concept of German studies has undergone major transformation. The traditional mixture of language and literary studies, related very closely to the discipline as practised in German universities, has expanded to embrace history, politics, economics and cultural studies. The conventional boundaries between all these disciplines have become increasingly blurred, a process which has been accelerated markedly since German unification in 1989/90.

New Perspectives in German Studies, developed in conjunction with the Institute for German Studies at the University of Birmingham, has been designed to respond precisely to this trend of the interdisciplinary approach to the study of German and to cater for the growing interest in Germany in the context of European integration. The books in this series will focus on the modern period, from 1750 to the present day.

Titles include:

Michael Butler and Robert Evans (*editors*)
THE CHALLENGE OF GERMAN CULTURE
Essays Presented to Wilfried van der Will

Michael Butler, Malcolm Pender and Joy Charnley (*editors*)
THE MAKING OF MODERN SWITZERLAND 1848–1998

Wolf-Dieter Eberwein and Karl Kaiser (*editors*)
GERMANY'S NEW FOREIGN POLICY
Decision-Making in an Interdependent World

Jonathan Grix
THE ROLE OF THE MASSES IN THE COLLAPSE OF THE GDR

Margarete Kohlenbach
WALTER BENJAMIN
Self-Reference and Religiosity

Henning Tewes
GERMANY, CIVILIAN POWER AND THE NEW EUROPE
Enlarging Nato and the European Union

Maiken Umbach
GERMAN FEDERALISM
Past, Present, Future

New Perspectives in German Studies
Series Standing Order ISBN 0–333–92430–4 hardcover
Series Standing Order ISBN 0–333–92434–7 paperback
(outside North America only)

You can receive future titles in this series as they are published by placing a standing order. Please contact your bookseller or, in case of difficulty, write to us at the address below with your name and address, the title of the series and the ISBN quoted above.

Customer Services Department, Macmillan Distribution Ltd, Houndmills, Basingstoke, Hampshire RG21 6XS, England

Walter Benjamin

Self-Reference and Religiosity

Margarete Kohlenbach
Lecturer in German and European Studies
University of Sussex

palgrave
macmillan

First published 2002 by
PALGRAVE MACMILLAN
Houndmills, Basingstoke, Hampshire RG21 6XS and
175 Fifth Avenue, New York, N. Y. 10010
Companies and representatives throughout the world

PALGRAVE MACMILLAN is the global academic imprint of the Palgrave Macmillan division of St. Martin's Press, LLC and of Palgrave Macmillan Ltd. Macmillan® is a registered trademark in the United States, United Kingdom and other countries. Palgrave is a registered trademark in the European Union and other countries.

ISBN 0–333–99359–4

This book is printed on paper suitable for recycling and made from fully managed and sustained forest sources.

A catalogue record for this book is available from the British Library.

Library of Congress Cataloging-in-Publication Data

Kohlenbach, Margarete.
Walter Benjamin: self-reference and religiosity/Margarete Kohlenbach.
p. cm. – (New perspectives in German studies)
Includes bibliographical references (p.) and index.
ISBN 0–333–99359–4 (cloth)
1. Benjamin, Walter, 1892–1940. 2. Religion–Philosophy–History–20th century. 3. Autopoiesis–History–20th century. I. Title. II. Series.
B3209.W584 .K64 2002
193–dc21

2002022419

10 9 8 7 6 5 4 3 2 1
11 10 09 08 07 06 05 04 03 02

Printed and bound in Great Britain by
Antony Rowe Ltd, Chippenham and Eastbourne

Contents

Acknowledgements

Sections 3.2.1 and 3.2.2 have been previously published in a shorter version, in Italian translation, as '*Mortificazione* e riflessione: Il Romanticismo nel saggio sul *Trauerspiel* di Walter Benjamin', in Barnaba Maj, Davide Messina (eds), *Walter Benjamin: tra critica romantica e critica del Romanticismo*, pp. 215–29, Aletheia: Firenza 2000, and are reproduced here by permission of Professor Barnaba Maj and Aletheia. Quotations from Walter Benjamin, *Gesammelte Schriften* are reproduced by permission of Suhrkamp Verlag.

The University of Sussex supported the completion of this book by granting me leave, both paid and unpaid; I thank especially my colleagues in the German Subject Group, who carried the main burden of these arrangements.

I received a great deal of support and encouragement from people who commented on the work in progress, invited me to present my thoughts at conferences and research colloquia or simply discussed Benjamin's work with me at great length. I wish to thank Jay Bernstein, Rüdiger Bittner, Cathriona Bourke, Raymond Geuss, Steve Giles, David Groiser, Catherine Hollis, Esther Leslie, Ladislaus Löb, Barnaba Maj, Nick Saul and Edward Timms. With admirable expertise and patience, Ladislaus Löb helped me in my efforts to present Benjamin's texts and my observations on them both accurately and in 'real' English.

I am grateful to my daughter Anna for letting me participate, while this book was being written, in her joyful discovery of human language.

Introduction

In 1938, Walter Benjamin published a short review of the work of the Institute for Social Research, the central body of the Frankfurt School. Towards its end, he raised the question whether the decay of liberal society left behind any elements worth saving for the future or, indeed, for mankind. The value which scholarly work might possess, he suggested, depended on a positive answer to this question. Benjamin did not give an answer, however, but concluded his presentation of Critical Theory by stating that the attempt to read a 'yes' from the lips of history was not of an academic kind.[1]

This book is an academic study. It uses the tools of conceptual and literary analysis to show that Benjamin's work diverges from academic writing, and how and why it does so. The book, that is to say, is deliberately uncongenial to its subject matter. Its aim is to encourage a recognition and study of Benjamin's work that is not blind to its basic, theoretically unsupported intentions. More often than not the literature on Benjamin combines a superficial empathy with his work with a lack of clarity about its character. The empathy is only superficial because it leaves the authors' presuppositions largely unchanged, be they neo-Marxist, poststructuralist, hermeneutical, positively religious or areligious. The lack of clarity is perhaps most evident in the confused use of such words as 'theory', 'philosophy', 'experience', 'theology', 'profane' or 'secularisation'. The statement that the present book is deliberately uncongenial to Benjamin's writing of course implies that I, like other commentators, do not abandon the non-Benjaminian assumptions which guide my interpretation. In my case, the most important assumption of this kind consists in the view that the conventions of traditional scholarship sometimes lead to a clarity of understanding which cannot be obtained by other means. In conceding uncongeniality, moreover, I renounce both any analytically dispensable empathy and any claim at presenting an 'immanent' critique. I do so in favour of the attempt to achieve clarity.

This attempt may be important. Benjamin's work represents one of the most radical and problematic twentieth-century responses to a dilemma that is still with us. The value of liberal culture cannot be demonstrated by the (liberal) exercise of rationality. The fact that this claim is either true, or at least, widely held to be true, shows that the re-establishment of liberal societies in the second half of the twentieth century may lack a solid basis. We therefore ought to understand and examine, and not merely adopt and commemorate, Benjamin's work.

Like many members of his generation, Benjamin sensed the problematic character of liberal culture long before its breakdown, in Germany, in 1933. From the early 1910s onwards, his response consisted in a particular appeal to the traditions of religious practice and thought. My central claim is that in his writing he pursued irreducibly religious objectives and did so, moreover, independently of any existing religious institution or positive religious belief. The latter qualification explains why in my title I speak of Benjamin's religiosity rather than of his religion. By describing his texts as a form of religious rhetoric, I distinguish them from (secular) theory and philosophy. This distinction is related to the traditional view that an argument differs in kind from a rhetorical plea, for instance a plea for revelation, redemption or justice. There is no obligation to engage with history or philosophy in an academic or philosophical manner, and rhetorical practices can be of existential importance. My use of the terms 'rhetoric' and 'religiosity' is not meant to belittle the moral commitment underlying Benjamin's work.

Benjamin expressed his simultaneous reliance on, and distance from, traditional religion in response to a letter by Max Horkheimer of 16 March 1937. In this letter, Horkheimer emphasises that the murders of the past cannot be rectified. The idea of such rectification, he argues, implies a belief in the Day of Judgement. In his sheaf 'N' from the Arcades project, Benjamin notes that

> history is not exclusively a human science; it is no less a form of remembrance. What science has 'substantiated', remembrance can modify. Remembrance can transform the incomplete (happiness) into something complete and the complete (suffering) into something incomplete. That is theology; but in remembrance we have an experience that forbids us to conceive of history on the basis of

> atheological principles, just as we must not try to write it by using directly theological concepts. [GS V.1, 588–9]

In spite of his unwillingness to use 'directly theological' concepts, Benjamin cannot renounce theological principles, for what he calls the 'experience' of remembrance is religious in kind. It may not actually imply the belief in any particular theological doctrine, as Horkheimer holds, but it certainly implies the general religious assumption that the life that was taken from those murdered was not their only and decisive one. Therefore a consistently areligious person will not accept that Benjamin's 'remembrance' denotes an experience on which any epistemological claims – for instance, on the manner in which we can, or ought to, write history – can be based. For such a person, the so-called experience consists in a belief, or in the interpretation of a feeling, which it is misleading to present as experience. To hold the belief in question, moreover, may or may not be valuable in psychic and political respects. The same is true of the opposite belief, of the recognition, that is, that in history there is irretrievable loss and irreparable injustice.

As we shall see, there is strong evidence for the view that Benjamin considered himself a religious writer in spite of his distance from traditional religions. The attempt or inclination to engage in, or inaugurate, non-traditional forms of religious practice and thought cannot be called 'secularisation' if this notion is to retain its meaning. I accordingly reject those interpretations of Benjamin's work that draw on the notion of secularisation. The view that 'secularisation' is unsuited to describe what happens in Benjamin's texts is relatively unproblematic with regard to his earliest and latest writings. The first pieces, still written in the context of the German Youth Movement, explicitly call for a new religion and try to provoke the revelation of an as yet unknown God; the latest notes urge a religious re-interpretation of Marxism and model what they term 'historical materialism' upon the tradition of *Heilsgeschichte*. The importance of religion in the writings between the First World War and 1940 is less evident. However, I shall defend the view that religious intentions are also essential to both the apparently academic and the most 'materialist' parts of Benjamin's work.

Max Weber criticised the post-traditional religiosity that flourished in German society around the First World War as self-deception.[2] Perhaps some readers will respond to my interpretation of Benjamin with a similar judgement. Quite apart from the fact that we may not really understand 'self-deception',[3] however, there are strong reasons to reconstruct Benjamin's work without resorting to this notion. First, there is sufficient textual evidence for the view that Benjamin knew exactly what he was doing. Second, what he believed he was doing was not to make himself believe what he knew to be false, but to try to believe in a meaning of history and life that he knew he could not prove. Third, it is unlikely that any person can afford to act exclusively on the basis of assumptions that can be shown to be correct. (It is sometimes reasonable to trust people even if we cannot be certain that they are trustworthy.) Finally, the notion of self-deception is closely related to that of inauthenticity, which figures prominently in the registers of modern anti-Semitism. This does not make anyone who considers a sceptical Jew's pursuit of religious fulfilment self-deceptive an anti-Semite, but it shows that in so doing we miss the chance of recognising, in what we condemn, a problem of *our* culture.[4]

The thematic focus of my discussion lies in what I call Benjamin's postulates of self-reference. Throughout the book, 'self-reference' designates a self-relatedness or *Selbstbeziehung* and is not meant to imply that the relation in question is one of 'reference' in any of the senses in which 'reference' is used in modern semantics. 'Self-reference' widely serves as an English equivalent of *Selbstbeziehung* and is perhaps more common and less awkward than 'self-relatedness' or 'self-relation'. The relatively vague character of 'self-reference', as I use the word, makes it a suitable umbrella term for Benjamin's various postulates, for he uses the structure of self-reference in a number of different contexts: the communication of language, the reflection of thought, the 'reflection' of the work of art, as well as the formation of modern technology and social organisms. Historically related to German Idealism and Early Romantic thought, the prominence and varied use of this structure in his work can strike readers who are unfamiliar with these traditions as an arbitrary complication. However, both the emphasis on and the variations of 'self-reference' are essential to the spiritual concerns of Benjaminian prose. To a very large extent, self-reference serves as a cipher – a non-sensual image, as Benjamin might have put it – of the absolute, or of a God who is no longer found in traditional religion. And given that he deems religion indispensable and traditional religions unconvincing, Benjamin is both free and impelled to search for religious fulfilment pretty much everywhere: in culture, art, literature and history, in the cinema and modern technology, in political events and new formations of matter; or in his own modes of writing, such as commentary, translation and historiography.

In Chapter 1, I examine Benjamin's search under the heading of 'language'. In his essay 'On Language as Such and on the Language of Man' (1916), Benjamin deals with the notion that every language communicates itself in itself. This notion ascribes to language in general a structure which the Judaeo-Christian and metaphysical traditions reserve for the absolute, the self-revealing deity or the divine 'word'. My analysis of Benjamin's related use of 'medium' and 'immediacy' ('Unmittelbarkeit') shows that magical beliefs, aesthetic or 'auratic' forms of experience and erotic desire all prove insufficient to explain his assertion of linguistic self-reference. 'Name', as I then argue, condenses this assertion into one word which, if we interpret it as part of a theory, is supported exclusively by the theological assumptions of Benjamin's Genesis commentary. Yet unlike the theological and metaphysical traditions on which he draws – notably medieval scholasticism and Johann Georg Hamann's theology of language – Benjamin's essay offers neither a positive theology nor a metaphysical or philosophical conception of language. As a whole, it does not even unambiguously maintain linguistic self-reference but realises what it also programmatically requires: an irreducible romantic wavering ('Schweben') about that identity of spirit and language which implicitly underlies its introduction of the linguistic absolute. I explain the simultaneous affirmation and negation of that identity, and thus of Benjamin's linguistic absolute itself, on the basis of his reflections upon his Genesis commentary. These reflections, as I try to show, express both the author's religious scepticism and his desire for apocalypse and divine revelation. Together, scepticism and desire permit us to understand that in the essay the 'anticipation' of apocalyptic fulfilment in the assertion of the linguistic absolute is also negated. It is the reference to a desire for religious fulfilment that distinguishes my interpretation from that of post-structuralist commentators, who were among the first to recognise the importance of 'self-reference' in Benjamin. Accordingly, I represent Benjamin's notion of linguistic self-reference not as part of a free-floating 'theory' or a culturally neutral deconstruction of Western metaphysics, but as the expression of a specific intention which, moreover, figures within a specific cultural context. Apart from the decline of traditional religions, two important features of early twentieth-century German society characterise this context: the conservative critique of culture, and the Youth Movement. With his 1916 postulate of linguistic self-reference Benjamin offers an esoteric re-interpretation of that highly spiritual understanding of the Youth Movement and culture in general which he developed under the guidance of his former teacher Gustav Wyneken.

Chapter 2 deals with the postulate of literary self-reflection in Benjamin's doctoral dissertation *The Concept of Art Criticism in German Romanticism* (1920). This study, sometimes regarded as the most academic of Benjamin's works, represents a particular adaptation of Fichtean philosophy and of the poetological thought of Friedrich Schlegel. It tries to show that in reflecting and criticising itself, the literary work of art dissolves into, and leads to, the absolute or the idea (of art). I analyse Benjamin's central claims, namely, that according to Fichte reflection is immediate knowledge, and that Schlegel considers the infinity of reflection to be 'intensive' or full. If we assess Benjamin's account of Fichte and Schlegel on the basis of philosophical or philological criteria, I argue, we have to consider it absurdly inadequate. To some extent, the contemporary influence of neo-Kantian anti-essentialism can explain Benjamin's presentation of Fichte and of an apparently aestheticist Romantic poetology. However, Benjamin turns against the anti-metaphysical intentions of neo-Kantian epistemological formalism in the very process of adopting it, and, in so doing, heeds the actual metaphysical and religious implications of both Romantic thought and his own messianism. Ultimately, it is not secular neo-Kantianism but the influence of religious decisionism – of the determination, that is, to live, write and respond to art and literature *as if* religion were true although we do not know it to be true – which explains the character of Benjamin's dissertation. While his use of 'reflection' diverges considerably from both Fichte's and Schlegel's, it is certainly congenial to the Romantic endeavours to benefit from a new religion. To a large extent, moreover, his conception of literary self-reflection conforms to what he plausibly discusses under the heading of (Schlegel's) 'mystical terminology'. Since such mystical terminology is essential to Benjamin's own critical practice, we can describe the latter as mystical.

In Chapter 3, I focus on a number of texts written between 1924 and Benjamin's death in 1940. I discuss the continuing importance of 'self-reference' and its application to (literary) history in the *Trauerspiel* study (completed in 1925). After attempting to clarify the character of Benjamin's alignment with Marxism and his idiosyncratic use of 'materialism', I deal with the two texts that have dominated the neo-Marxist reception of Benjamin. I try to show that the postulate of literary self-reflection, albeit no longer explicitly upheld, underlies Benjamin's argument in 'The Author as Producer' (1934), and that his use of 'Technik' ('technique' *and* 'technology') agrees with the related practice of 'mystical terminology'. Accordingly, I interpret Benjamin's engagement with Brechtian theatre as still shaped by those religious intentions that informed the explicit discussion of literary self-reflection in the dissertation. The representation of cinema and of apparently non-auratic forms of reception in 'The Work of Art in the Age of Its Technical Reproducibility' (1936), as I then argue, presupposes again the irrelevance of the subject–object opposition that Benjamin discards with his postulates of self-reference. It actually amounts to an esoteric affirmation of an emphatic, 'bodily', notion of contemplation or spiritual practice, which is in agreement with his attempt, at least from the early 1920s onwards, to redefine 'politics' in spiritual terms. This attempt is essential to Benjamin's notion of *Geistesgegenwart* ('presence of mind') as it informs his later constructions of both film reception and 'historical materialism'. Benjamin maintained that his last important text, 'On the Concept of History' (1940), contained a methodological summary of his former work. In conclusion, I try to show that his religious rhetoric of self-reference can be considered the 'method' in question.

Benjamin attributed to National Socialism an aestheticisation of politics and countered it with what he called the politicisation of aesthetics. To the same extent that his work is adequately described as an attempt to respiritualise modern politics and culture, it can appear uncannily close to those features of National Socialism which are currently discussed under the heading of 'political religion'. In my view, we abandon the enlightened intentions that were part of the reception in the 1960s and 1970s if we simply refuse to take this impression seriously. It is, of course, problematic to acknowledge any affinities between Benjamin's work and that radical German conservatism which paved the way for, and partly merged with, Nazism. We are mistaken, however, in responding with surprise or disbelief to similar cultural orientations on both sides of the divide that irrevocably separates the victims from the murderers. Such a response is mistaken precisely because the Nazis' characterisation of German Jewry as alien to German culture was arbitrary. To recognise similarities, moreover, is not to identify. While Benjamin's alignment with Marxism is certainly an expression of the difference of his position from that of non-Jewish German conservatives, it may be too ambiguous to define what distinguishes his *thought* from radical conservatism. As far as I can see, there are at least two indisputable and clear differences: firstly, Benjamin's life-long refusal to associate his hope for a spiritual revival with any variety of nationalism; and secondly, his related anarchic messianism, which, whenever it became dominant in his writing, ruled out support for any stable form of social or political order. Ironically, it is precisely his remoteness from secular political intentions, which distinguishes his thought from that of those contemporaries who re-introduced magico-religious hierarchies into modern society. As soon as we construe the expression of his desire for religion not as a form of spiritual exercise but as a political programme or 'theory', we ourselves blur that difference. Benjamin's sporadic attempts to shed his contemplative self are problematic not because contemplation is preferable to practice, but because any political practice which aims for spiritual fulfilment is fraught with danger.

References and Abbreviations

Primary and secondary sources are first cited by their complete title, then frequently with a characteristic abbreviation. Arabic numerals usually refer to the pages, Roman numerals to the volume number or the preliminary pages, of a work. English quotations from the Bible are taken from *The Bible: Authorized King James Version*, eds Robert Carroll, Stephen Prickett, Oxford 1997, German ones from *Die gantze Heilige Schrifft*, trans. Martin Luther, ed. Hans Volz, 2 vols, Munich 1972 (= repr. of the first edition: Wittenberg 1545). Unless otherwise indicated, all translations are mine. The following abbreviations are used:

AB	Theodor W. Adorno, Walter Benjamin, *Briefwechsel 1928–1940* (= Theodor W. Adorno, *Briefe und Briefwechsel*, vol. I), ed. Theodor W. Adorno Archiv, Frankfurt/M. 1994.
DW	Jacob and Wilhelm Grimm, *Deutsches Wörterbuch*, 32 vols, Munich 1984 (= repr. of the edition: Leipzig 1854–1960).
GB	Walter Benjamin, *Gesammelte Briefe*, ed. Theodor Adorno Archiv, 6 vols, Frankfurt/M. 1995–2000.
GS	Walter Benjamin, *Gesammelte Schriften*, eds Rolf Tiedemann, Hermann Schweppenhäuser, 7 vols, Frankfurt/M. 1974–1989.
HWP	Joachim Ritter, Karlfried Gründer (eds), *Historisches Wörterbuch der Philosophie*, 12 vols, Darmstadt 1971–.
KA	Friedrich Schlegel, *Kritische Friedrich-Schlegel-Ausgabe*, ed. Ernst Behler, 35 vols, Paderborn 1958–.
SW	Walter Benjamin, *Selected Writings*, ed. Michael W. Jennings, 3 vols, Cambridge (Mass.)1996–.

Quotations from Benjamin's texts are taken from GS and translated by myself. Additionally, I refer to the corresponding passages in SW I and SW II. In Chapter 1, mere page numbers relate to 'On Language as Such and on the Language of Man' in GS II.1, mere page numbers in italics to the corresponding passages in SW I. In Chapter 2, the same procedure is used with regard to *The Concept of Art Criticism in German Romanticism* in GS I.1 and SW I.

1
Language

According to the most comprehensive characterisation of language in 'On Language as Such and on the Language of Man', language is, in every case, 'not only communication of the communicable but also, at the same time, a symbol of the non-communicable' (156; *74*). The central claim concerning language as communication is what I call Benjamin's thesis of linguistic self-communication or (L).

(L) Jede Sprache teilt sich *in* sich selbst mit, sie ist im reinsten Sinne das 'Medium' der Mitteilung. [142; *64*]

(L) Every language communicates itself *in* itself, it is the 'medium' of communication in the purest sense.

As to the symbolic function of language, Benjamin states that it is not considered in the essay, at least not explicitly (156; *74*). He thus indicates that the notion of a linguistic symbolisation of something non-communicable may be of *implicit* importance in 'On Language'. This is indeed the case. The view that language is somehow related to what is non-communicable finds its tacit expression in the ambiguous way in which Benjamin treats the ideal of perfect, self-contained communication that he erects with (L).

(L) presupposes three possibly nonsensical assumptions. It presupposes, first, that language is the communicator in, or the subject of, communication, second, that the object and the medium of communication are identical, and third, that the subject and the object of communication – the communicator, that is, and that which is communicated – are identical, too. As a result, language counts as self-contained communication. Such self-containment seems to rule out the notion that language is essentially related to something other than itself, for example something uncommunicated or even non-communicable. However, Benjamin does not unambiguously maintain the three assumptions underlying (L) throughout the essay. In his initial discussion of spiritual self-expression, he either conceives an alternative to, or expresses strong reservations about, each of them. And the simile in the concluding paragraph of the essay draws on a notion of spiritual self-expression that conflicts with (L). While comparing language to a secret password,[1] Benjamin implies that it is an entity different from language, which communicates, for it is the sentry who speaks the password, not the password itself or the sentry's language. This means that a twofold darkness pertains to 'linguistic self-communication'; it is not only the meaning of (L) that is difficult to determine but also the status which (L) has in the essay. As we shall see, Benjamin does not explain the problematic meaning of 'linguistic self-communication'. And though maintaining (L), he allows for the possibility of its incorrectness. I shall approach his experiment with (L) by considering spiritual self-expression first.

1.1 Spiritual self-expression

Benjamin bases his notion of spiritual self-expression on a very broad understanding of 'language'. There is no event or entity, neither in animate nor in inanimate nature, which does not partake of language, he claims, and 'language', as used here, is by no means a metaphor (140–1; *62*). An entity partakes of language by communicating its spiritual content. The spiritual content that is communicated in a particular language is the spiritual essence ('das geistige Wesen') of the communicating entity (142, 157; *63, 74*). Nothing is devoid of spiritual essence since each entity communicates the spiritual essence belonging to it. Such communications form a continuous stream that runs through everything there is, from the lowest being to man, and from man to God (157; *74*). The unity of this movement is the divine word of creation, a word that inheres, though in a modified manner, in the spiritual essences of both man and nature. Since to partake of the stream of communication means to partake of language, 'language' designates a sphere or mode in which the divine and the non-divine meet. Like similar notions in (neo-)Platonic thought, Benjamin's 'language' designates a kind of communion that exists between God, or the supreme idea, and all beings.

The higher or lower degree of consciousness present in the various occurrences of communication does not affect their basic linguistic quality (141; *62*). There is communication between natural objects, which occurs in a material communion ('stoffliche Gemeinschaft') (147; *67*). The communion between man and nature is spiritual, and man's verbal language merely a particular instance of that communication which pervades the universe (140; *62*). The term 'communication' in the notion of a material *communicatio realis* and in that of a spiritual *communicatio verbalis*,[2] we may express Benjamin's view, is not an equivocation; *communicatio realis* and *communicatio verbalis* share *one* linguistic nature. Accordingly, he uses 'language' to denote what spiritual and material kinds of impartation have in common. This usage is related to the idea, prominent for example in Romanticism, that spirit ('Geist') and material energies spread in a similar, if not identical, way.[3] Only, the notion of similarity does not suffice for Benjamin's purposes since he claims his use of 'language' to be unmetaphorical. Judged by the standards of twentieth-century ordinary usage, this claim is wrong. What underlies Benjamin's broad understanding of 'language', however, is an epistemological assumption for which the facts of empirical linguistics are, or are taken to be, immaterial. In accordance with the contemporary discussion of a 'material', or non-subjective, configuration of the *form* of knowledge,[4] he assumes that we can conceive an object only if it expresses and communicates ('im Ausdruck mitteilt') its spiritual essence (141; *62*). If man's verbal language is an instance of the expressiveness in everything knowable, moreover, it cannot be a metaphorical model for that expressiveness. These assumptions, of course, should not be taken for granted. In his 'On the Programme of the Coming Philosophy' (1917), Benjamin envisaged, but did not elaborate, a corresponding epistemology. And as will become clear in what follows, even verbal language is actually not, at least not exclusively, an expression of spiritual essence in Benjamin's sense.

How, then, does Benjamin conceive the expression of spiritual essence? Language, he writes, communicates the spiritual essence corresponding to it (142; *63*). Spiritual essence is that which is communicated, and language seems to have the communicating part. However, the next sentence reads as follows: 'It is of fundamental importance to know that this spiritual essence communicates itself *in* language and not *through* language.' Let us disregard for the time being the distinction between communication *in* and communication *through* and focus on the demonstrative pronoun 'this' before 'spiritual essence'. It implies that 'spiritual essence' refers to the same thing as in the preceding sentence. But now spiritual essence, and not language, counts as the subject of communication. What is more, spiritual essence is now said to communicate in language, it is said to communicate, that is, in what in the preceding statement appeared to be the subject of communication. The two statements can be made compatible by taking the 'language' of the first as a mere grammatical subject, with the activity in question, communication, being exercised by something else, call it the 'actual subject'. The first statement then means that a spiritual essence is communicated in language, and the second statement specifies the (actual) subject of this communication as a spiritual essence. This interpretation accords with German (and English) usage. Luther, for instance, uses 'deine sprache verrhet dich' ('thy speech bewrayeth thee', Matt. 26: 73) to express the idea that Peter betrays himself by talking the way he does.

The interpretation, moreover, accords with Benjamin's distinction between language and spiritual essence: 'The German language, for example, is by no means the expression of all that which we can express *through* it – or so it is thought – but the immediate expression of that which communicates *itself* in it. This 'itself' is a spiritual essence. For the time being, it is therefore obvious that the spiritual essence which communicates itself in language is not language itself but something that must be distinguished from it' (141; *63*). In the light of this passage, Benjamin's statement that language communicates the spiritual essence corresponding to it not only can but must refer to the self-communication of a spiritual essence in language.

The example of the German language calls to mind the various theories, Humboldt's among them, of a national spirit which expresses itself in the language of the nation concerned.[5] In contrast to what Benjamin rejects as the 'bourgeois' ('bürgerlich') view of language, (143–4, 150; *64–5*, *69*), these theories do not consider language a neutral means for the description of facts but a medium which colours whatever may be described in accordance with the spirit, generating force or principle of that language. The idea of language being a medium of self-expression is also related to the notion of style.[6] To illustrate Benjamin's thought, one might think of, say, the vigorous style in which the vigorous character of a speaker is expressed. The spiritual essence that a language is said to communicate relates to that language in the same way as a vigorous person or her vigour relate to the characteristic style she uses.

The claim that language communicates the spiritual essence corresponding to it thus relies on the idea of a subject's, or a spirit's, self-expression. That is why the term 'Wesen' in Benjamin's essay frequently and strongly evokes a sense of the German 'Wesen' which is not covered by the English 'essence', namely 'animate being'. For example, by answering the question 'To whom does he [man] communicate himself?' by 'Man's spiritual essence communicates itself to God' (143–4; *64–5*), Benjamin comes close to using 'man's spiritual essence' as an equivalent of 'man'. In accordance with his intention to free post-Kantian epistemology from the doctrine of constitutive subjectivity,[7] however, he also tries to minimise his reliance on the idea of a spiritual being or personal communicator. We already saw that he uses 'language' as a grammatical subject where he actually considers a spiritual being the subject of communication; and in the statement that 'spiritual essence [...] is taken to consist precisely *in* communicability' (145; *66*), 'spiritual essence' refers not to any being but to the essence of spirit. The word 'Wesen', meaning both 'essence' and 'animate being', is ideally suited for Benjamin's purposes. In the latter sense it allows him to draw, in combination with 'spiritual', on the idea of a personal communicator, while in the former sense it enables him to suggest that he does not draw on that idea. Thus both his emphasis on the subjective domain of self-expression and his rejection of a subject-oriented epistemology can assume an appearance of plausibility.[8]

In maintaining that language communicates the spiritual essence corresponding to it, then, Benjamin introduces a communication in language that is not a self-communication of language. He calls it obvious that the spiritual essence that communicates itself in language is not language itself, and obvious it certainly is. The national spirit that supposedly informs the German language is not that language itself, and to say that vigour or a vigorous person express themselves in a vigorous style is not to say that a vigorous style expresses itself in a vigorous style. The medium in which someone or something communicates is not at the same time the communicator or subject of that communication. We should note, however, that Benjamin calls the difference between the subject and the medium of communication obvious only 'for the time being' ('zunächst') (141; *63*). This reservation is meant to leave the door open for his later introduction of linguistic self-communication.

In his discussion of the self-communication in language of a spiritual essence, Benjamin distinguishes not only between the communicating subject and its medium, but also between that subject and the object of communication, which he considers the linguistic essence ('sprachliches Wesen') of an entity. Language, he argues, 'communicates the linguistic essence of the corresponding entities, their spiritual essence, however, only in so far as it is immediately contained in their linguistic essence, only in so far as it is communic*able*' (142; *63*). Here, the phrase 'linguistic essence' can be interpreted on analogy of the expression 'picturesque landscape'. A landscape that lends itself to being painted is picturesque, and that part or aspect of an entity that lends itself to being expressed in language is its linguistic essence. Thus the linguistic essence of an entity is not language, just as the picturesque character of a landscape is not painting.[9] Given this notion of linguistic essence, what Benjamin seems to say is that spiritual essence does not communicate itself but *something of* itself, namely that which lends itself to being communicated and which, by consequence, is communicable. Assuming that not everything in a spiritual essence is communicable, it follows that Benjamin's notion of spiritual communication is incompatible with that of 'self-communication' in the strict sense; for the subject and the object of spiritual communication cannot be identical if not everything in the subject is communicable.

But does Benjamin assume that spiritual essence cannot be communicated in its entirety? In the passage under discussion, he suggests a partial incommunicability of spirit by explicitly denying the 'external' identity of spiritual and linguistic essence, of spiritual essence, that is, and that which is communicable of it (142; *63*). If, as his denial seems to imply, 'spiritual essence' and 'linguistic essence' are not co-extensive terms, spiritual essence must comprise something incommunicable. To be sure, Benjamin's wording suggests that another, presumably 'internal', kind of identity exists between the two. Yet if this identity were to amount to the unrestricted communicability of spiritual essence, his distinction between spiritual and linguistic essence would be empty and superfluous. I therefore assume that he proposes, *in the context of 'spiritual self-expression'*, a partial incommunicability in spirit and, by extension, the notion that the subject and the object of communication are not identical.[10] We should note, however, that he phrases his distinction in such a way that an identity of that which is distinguished is not excluded. By stating that spiritual and linguistic essence are identical only 'in so far' as the former is communicable, Benjamin does not rule out that they are identical, period; for spiritual essence may be entirely communicable after all. Again, he tries to leave the door open for linguistic self-communication to enter the scene.

To sum up, independently of his wish to introduce (L), Benjamin's notion of spiritual self-expression does not present major problems of comprehension. It does not make language the subject of communication, nor does it imply the identity of either the subject and the medium, or the subject and the object, of communication. Spiritual self-expression agrees with the following model: Whatever she may mean to say, vigorous Lucy communicates her vigour by speaking in a vigorous way. Unlike Benjamin, we may hesitate to ascribe this kind of communication to events and inanimate entities, but we certainly understand what it is that is being ascribed.

1.2 The self-communication of language

1.2.1 Introducing linguistic self-communication

After the distinction between spiritual and linguistic essence, Benjamin introduces his thesis of linguistic self-communication (142; *63–4*). Before stating (L), he twice maintains what I consider its curtailed variant (L*): 'Every language communicates itself'. (L) follows almost immediately upon the second statement of (L*), which in turn is preceded by the claim (L**): 'A spiritual essence communicates itself *in* what is communicable *of* it'.

The reasoning that leads to the first statement of (L*) is the following: 'Language communicates the linguistic essence of an entity. However, the clearest manifestation of that essence is language itself. The answer to the question "*What* does language communicate?" is therefore "*Every language communicates itself*".' Benjamin considers this passage a transition, and plausibly so. The first sentence can be read as a summary of the preceding discussion. It then means that the linguistic essence of an entity is communicated in language, and we can add that it is a spiritual essence that does the communicating. With the two subsequent sentences, Benjamin introduces something new, namely (L*). (L*) is new because it differs from both the first sentence of the present passage and the preceding discussion by not allowing the assumption that the subject of communication is different from language.

If (L*) implies that the subject and the object of communication are the same, the 'itself' of the object position is an equivalent of 'language'. In the search for a subject different from language, we may transform (L*) into 'In every language, this very language is communicated'. However, if we then add an actual subject different from language – as the context urged us to do in the case of 'Language communicates the spiritual essence corresponding to it' – we no longer interpret (L*), but change its meaning. In 'In every language, this very language is communicated by x', with x being different from 'language', any identity between the subject and the object of communication is irretrievably lost. Of course, (L*) as such does not necessarily imply that the subject and the object of communication are identical. On the analogy of expressions such as 'Lucy expresses herself', we may take (L*) to mean that every language communicates not itself, but something of itself, say, an attribute or a part. The reflexive verbal construction in (L*) then stands for the identity of the subject of communication with that which contains the communicated attribute or part, that is, with language. This again means that we cannot, without changing its meaning, transform (L*) along the lines of the transformation which did work for 'Language communicates the spiritual essence corresponding to it'.

With (L*) then, Benjamin claims language to be the actual subject of communication. The inference, however, on which he bases the transition to (L*) is incorrect. (L*) does not follow from the assumption that the clearest manifestation of linguistic essence is language. For language to be communicated if linguistic essence is communicated, language must *be* linguistic essence, and not only its clearest manifestation. Benjamin is aware of this. Before stating (L*) for the second time, he dismisses the notion of manifestation and claims the identity of linguistic essence and language, the identity, that is, of what he had introduced as the object and the medium of communication respectively. It is this equation on which he relies when stating (L*) for the second time. Since the second statement of (L*) precedes the assertion of linguistic self-communication in its complete version (L), it is not surprising that he stresses the importance of adequately interpreting the equation of linguistic essence with language: '*The linguistic essence of an entity is its language*. The understanding of linguistic theory depends on carrying this proposition to a level of clarity at which even the appearance of tautology is annihilated. [...] Everything rests on this "is" (equivalent to "is immediately").' I shall discuss Benjamin's equation of linguistic essence with language in the following section 1.2.2.

What matters here is that the impact of the equation is smaller than Benjamin's emphasis suggests. As regards 'linguistic self-communication', it is not the case that *everything* rests upon the 'is' of the equation, for the identity of linguistic essence and language is not a sufficient condition for the truth of either (L*) or (L). What eventually leads Benjamin to state (L) is the following reasoning:

(L**) Was *an* einem geistigen Wesen mitteilbar ist, *in* dem teilt es sich mit;
das heißt:
(L*) jede Sprache teilt sich selbst mit.
Oder genauer:
(L) jede Sprache teilt sich *in* sich selbst mit, sie ist im reinsten Sinne das 'Medium' der Mitteilung. [142; *64*]

(L**) A spiritual essence communicates itself *in* what is communicable *of* it;
that is to say,
(L*) every language communicates itself.
Or more precisely,
(L) every language communicates itself *in* itself, it is the 'medium' of communication in the purest sense.

In (L**), Benjamin puts his equation of linguistic essence with language to use by placing the object of communication, i.e. that which is communicable of a spiritual essence, in the position of the medium of communication. With the phrase 'that is to say' he then pretends (L*) to be closely related, if not equivalent, to (L**). If this were the case, the step from to (L*) to (L) would be a small one, for if we disregard the terminological comment in the second clause of (L), (L) only combines the idea, of (L**), of a communication in the object of communication with the self-communication of language, stated already in (L*). Yet the transition from (L**) to (L*) is questionable since (L**) clearly identifies a spiritual essence as the subject of communication and (L*), as argued above, excludes any communicator different from language. So does (L), and for the same reasons. This means that (L*) and (L) are only compatible with (L**) if spiritual essence and language (or linguistic essence) are the same. Linguistic self-communication, as introduced by Benjamin, presupposes the identity of spirit and language. It thus presupposes what underlies his conception of name, which I shall discuss in section 1.2.3.

1.2.2 'Medium' and magic

Let us consider the equation of linguistic essence and language first. Benjamin comments on it as follows: '*The linguistic essence of an entity is its language*. [...] This sentence is non-tautological, for it means that that which is communicable of a spiritual essence *is* its language. ... that which is communic*able* of a spiritual essence is immediately language itself. Or: the language of a spiritual essence is immediately that which is communicable of it' (142; *63–4*). These comments contain two problems. 1. Why should there be any need to stress the absence of tautology? 2. What is involved in Benjamin's insistence on immediacy?

1. With his denial of tautology, Benjamin claims that the equation of linguistic essence with language represents a substantive insight. He underlines the denial by using the very strong words that I have already quoted: even the appearance of tautology is to be 'annihilated'. His anxiousness to eradicate the shallowness of tautology makes sense if we consider how he arrived at the equation. In his presentation, the equation is still part of his commentary on (L*), which he, on the occasion of its first statement, based on the assumption that language is the clearest manifestation of an entity's linguistic essence. And given that linguistic essence is that aspect of a spiritual essence which lends itself to being communicated, which lends itself, that is, to being expressed in language, the information that linguistic essence manifests itself most clearly in language is indeed not of breath-taking value.

Once any association with the notion of manifestation is discarded, however, the equation of linguistic essence and language clearly ceases to be threatened by tautology. As Benjamin convincingly maintains, the equation means that that which is communicable of a spiritual essence is its language and vice versa. There would be no possible object of spiritual communication, then, other than language. The picturesque, if communicated (or painted), '*is*' painting after all, and vigorous Lucy does not communicate her vigour by speaking in a vigorous way; all she can communicate '*in*' speaking in a vigorous way is her vigorous way of speaking. The problem with the equation of linguistic essence and language is not that it comes close to being a truism; the problem is that it seems blatantly false.

2. In the clause following 'for it means', Benjamin replaces the phrase 'the linguistic essence of an entity' with his previous characterisation of linguistic essence as 'that which is communicable of a spiritual essence'. This replacement to some extent specifies the meaning of the equation but provides no further explanation. The denial of tautology, as we saw, is misleading. Nor does the rejection of 'manifestation' provide an explanation since by dismissing 'manifestation', Benjamin merely catches up with one of the implications of the equation; of course, if language is the linguistic essence of an entity, it is not the manifestation of that essence. This means that 'immediacy' represents the only remaining option for a more than technical explanation of the view that linguistic essence and language are the same.

Benjamin uses the adverb 'immediately' three times in three consecutive sentences to qualify the identity of linguistic essence and language. Yet it is questionable whether 'immediate' is at all a sensible qualification of identity. If the morning star is the evening star, one may object, it is that neither immediately nor mediately, it just is the evening star. What can sensibly be called immediate or not is the way in which we gain knowledge of, or experience, an identity, and not the experienced identity itself. In the light of this objection, Benjamin's insistence on immediacy can be taken to indicate that he, in some way, is concerned with the experience of language and linguistic essence. I shall try to show in due course that the equation of linguistic essence and language can indeed be understood as being related to a number of particular experiences of communication. Such an explanation, however, clearly runs counter to Benjamin's ambitions. The author of 'On Language' sees himself as contributing to the metaphysics of language (146; *66*). He aims at saying what language is, and not how it is experienced, let alone under particular circumstances. To understand how Benjamin may explain the equation of linguistic essence and language, we cannot therefore draw on the notion of experience (in any non-religious sense).[11]

Talk of an immediate or mediated identity is usual practice in post-Kantian idealism. In claiming that essence is appearance, Hegel, for instance, takes the identity in question as being constituted through the mediation of a twofold negation.[12] Roughly, the identity of essence and appearance allegedly lies in the fact that 'essence' can only be specified as that which is not non-essential. Appearance, that is the non-essential, is essential to 'essence'; it '*is*' essence, or the truth about it. Benjamin's equation of linguistic essence and language comes close to suggest a 'mediated' identity of this kind. Language, after all, counted just a minute ago as the manifestation or appearance of (communicable) essence. Placed against this background, Benjamin's emphasis on immediacy is historically significant. It shows that he rejects an understanding of the identity of linguistic essence and language on the dialectical model. Hegel's 'essence is appearance' is still attractive, but the confidence that dialectical mediation will do the job of excavating essence from the depth of what is hidden is gone. However, the proviso that we ought not to understand the identity of linguistic essence and language along dialectical lines tells us very little about how we can understand it. The explicative value of Benjamin's recourse to 'immediacy' is extremely limited.

This limitation notwithstanding, Benjamin bases his notion of medium on 'immediacy'. In striking contrast to theories which consider language a medium because it is a system of signs, or means, which are employed to signify what they are not, he holds that what makes a medium – 'das Mediale' – is the immediacy of spiritual communication (142; *64*). A later passage of the essay confirms that the immediacy that the author considers essential to 'medium' is that of the identity of linguistic essence and language. What is communicated in the medium, he writes, 'is – in accordance with the medial relation – this very medium (language)' (145; *66*). Benjamin's German expression 'das mediale Verhältnis' is as unusual as the English 'the medial relation', especially if we – in line with his distinction between 'medium' and 'means' (144–5; *65*) – disregard the relation between means and ends. The second component that is required to make the notion of a relation meaningful seems to be lacking. Benjamin's use of the phrase, however, clearly shows that for him a medial relation exists if the object of a communication is its medium. The apparently missing second component is the object of communication and the relation between that object and the medium is one of identity.

What according to Benjamin makes a medium is the immediacy by which the object of communication is the medium itself (and vice versa). This allows us to interpret the second clause of (L). Language is a medium, for what is communicated in it is immediately language itself. And language is a medium in the purest sense, for Benjamin's notion of medium coincides with the purest sense of that word.

However, this 'purest sense' of 'medium' may actually fail to exist. Benjamin's use of the term diverges from ordinary and academic practice *and* is in danger of being paradoxical. It excludes the literal sense of 'medium', 'means'. It does not coincide with an understanding of 'medium' as a mean, situated for instance in the middle between the extremes of subject and object, for Benjamin's 'medium' is – and is undialectically – the object or extreme. The spiritualistic sense of 'medium' is excluded since the go-between between a spirit and spiritualistically ungifted humans is 'the medium' but not at the same time the object of communication. More importantly perhaps, Benjamin's notion of medium also excludes any sense of that word which allows us to say without contradiction that something other than the medium itself is communicated in a medium. And indeed, it may exclude any understanding of 'medium' that is compatible with the notion that a medium is at all that in which something is communicated. If the identity of the object and the medium, as Benjamin convincingly assumes, precludes the former from being manifested in the latter, it may also preclude it from being communicated in the latter. The equating copula may well rule out the relation of 'communication in' between what it equates. If Benjamin's notion of medium, in spite of being based on the identity of object and medium, is also to complement expressions such as 'to communicate in ...', it is therefore at least in danger of being paradoxical. And Benjamin does continue to speak of the communication in a medium after having equated the medium with the object of communication.[13]

With his notion of medium, then, Benjamin proposes an unusual and possibly paradoxical understanding of the term, which stands in need of argumentative support.[14] The declaration that the envisaged sense of 'medium' is the purest, promising though it may sound, does not do. Elsewhere, Benjamin uses the word 'pure' to qualify that knowledge the quintessence of which he calls God (GS II.1, 163; SW I, 104). We may therefore take the use of 'pure' in the second clause of (L) as an indication that the envisaged sense of medium is a theological one. In section 1.2.3 I shall argue that if 'On Language' provides any argumentative support for Benjamin's notion of medium, such support has indeed to be considered theological. Here, I am only concerned with the analysis of 'medium' on the basis of Benjamin's untheological introduction of the term.

In the immediacy of spiritual communication Benjamin does not see a problem related to his, so far at least, implausible use of 'medium', but the fundamental problem of all linguistic theory, and even of language itself (142–3; *64*). He concedes that 'one' may choose to call such immediacy magical.[15] This guarded way of introducing the vocabulary of magic contrasts sharply with the great importance of 'magic' for his argument. If there is a magic of language, and if Benjamin's notion of medium captures that magic, then his use of 'medium' is no longer implausible but adequate in view of an arational character of language which 'bourgeois' theories ignore. Because of its great impact, the notion of magic deserves scrutiny in spite of its guarded introduction. On what grounds can the immediacy of spiritual communication, as maintained in 'On Language', be considered magical?

Benjamin's notion of medium resembles a belief about representation that seems to be part of certain magical practices. When a magician mutilates an image in order to harm the person that the image represents, we may explain his action by the assumption that he believes the image to be the person. In more general terms, we may say that he believes that the medium of representation is not a means of representing an object, but the represented object itself. In this account of magical practice, the magician's belief is in structural agreement with that feature of spiritual communication which Benjamin expresses by claiming that language is linguistic essence. Furthermore, if we accept Benjamin's use of 'immediately' as a qualification of 'is', we may say that the magician considers the image to be immediately the person. At least the magician, in contrast to, say, the astronomer who believes the morning star to be the evening star, is unlikely to arrive at his belief through a process of observation, calculation and inference. On account of these similarities one may indeed call the immediacy of spiritual communication magical, just as Benjamin concedes.

However, there are strong reasons against calling it such. Magicians know that magic fails at times. Their recognition of possible failure reveals the limitations of the preceding account. If our magician's enemy goes unhurt in spite of the mutilated image, the magician will not normally respond to his lack of success by denying it. He will not normally say, that is, that since the person *is* the image, he actually harmed the enemy and all is fine. On the contrary, he is likely to repeat his magical assault, perhaps with a more thorough manipulation of the image and with more respect for the possible workings of adverse spirits. Although we may explain the destruction of the image with reference to the belief that the image is the person, the magician is far from holding throughout that the person is the image. There is no equivalent in what he believes to Benjamin's claim that linguistic essence is language. In using the image to harm the enemy, the magician is concerned with achieving an end that he does not consider identical with what he achieves in relation to the image (or means). He wants to harm the enemy and not the image. Only if we dissociate the destruction of the image from its embedding in purpose-oriented practice, can we be satisfied with describing him as someone who believes the image to be the object of representation. Although the magician is not concerned with scientific investigation, moreover, we ought to know better than to describe him as believing that the image is immediately the person; for this description makes him look a fool. It is after all the magician himself who manipulates the image in order to invest it with magical powers. He would be stupid to do so if he did not know that the image as such does not do the job (and that he would lose his if it did). Only if we disregard his manipulation and craft in general – perhaps because we consider them inappropriate anyway or because we are not interested in practical matters – can we take him to believe that the image is immediately the person.

The gap between magical attitudes towards representation and

Benjamin's notion of immediate communication widens if we take the fact into account that Benjamin rules out any object of communication other than language. The believer in magic, to the extent that he is interested in having an impact on events, relates to the world not as a theatre of communication but as a place for action. If his spirits were to persist in displaying nothing but their own communicative potentials, he might as well abandon magic altogether. Even sophisticated believers in magic such as Goethe's Faust are interested agents in this sense; they put their spirits to work. And even in fictional models of a magical world whose fairy-tale air seems to place them far above the morass of instrumentality, magical communication does have a content other than language. The mirror tells Snow White's step-mother that she does not match the girl's beauty. This information is of great importance to the characters' actions and fate. If we take the mirror to communicate exclusively its linguistic essence or language, we cease to understand the tale's magical world.

There is no significant similarity, then, between Benjamin's notion of immediate communication and magic as such. There are some similarities between that notion and a particular interpretation of magic. In construing the content of magical beliefs in dissociation from the concerns of purposive action,[16] that interpretation is questionable on general grounds. What matters in the present context is that Benjamin's reliance on a non-practical view of magic begs the question of the suggested superiority of his notion of medium over 'bourgeois' theories of language. If, as I argued, spiritual communication in Benjamin's sense can only be considered magical if one presupposes the irrelevance, to magic, of purposive action, the characterisation of language as magical also merely presupposes the irrelevance of the means–ends relation to linguistic theory. Benjamin's recourse to the magic of language is rhetorical. It is apt to suggest that his notion of medium overcomes the alleged emptiness of the 'bourgeois' approach by the recognition of a deep mystery, essential to language. Yet it does not do anything to show that his problematic notion of medium is tenable, tenable, that is, in contrast to the alleged untenability of 'bourgeois' theories (144; *65*). We learn from the author neither why his notion of medium should be considered true nor, indeed, why it should be considered at all.

So we must look for ourselves. Benjamin presupposes language to be a mystical and inexplicable reality.[17] Therefore any (objective) reason which we may wish to add in support of his theory of language is not in fact a reason for his theory of language. All we can hope to explain is why Benjamin wants to uphold that theory and its central notion of medium in spite of their incomprehensibility. What is the attraction of maintaining that linguistic essence is language, and is so immediately? Under what conditions does the contention that language is the exclusive object of communication assume an appearance of truth?

As we saw, the equation of linguistic essence with language replaces Benjamin's initial model of manifestation. While moving towards the identity option, he draws the distinction between the linguistic essence of a lamp and 'the lamp itself' (142; *63*). In his notes, the phrase 'the lamp in and for itself' indicates his concern with Kant's concept of the thing-in-itself (GS VII.2, 788). Kant's things-in-themselves remain unattainable to human knowledge in spite of their manifestation or appearance. In claiming that linguistic essence *is* language, Benjamin tries to preclude such frustrating elusiveness. And he tries to preclude it with regard to what matters most: essence. To be sure, the worry that 'manifestation' precludes the attainability of essence is out of place with regard to linguistic essence. 'Linguistic essence', after all, was introduced as that which is communicable of spiritual essence. Yet as a specification of 'essence', and with regard to the possible elusiveness of essence, that introduction merely represents a terminological decision. It is therefore not surprising that the Kantian spectre of perpetual elusiveness continues to haunt Benjamin even with regard to linguistic essence. The identity claim is his attempt at exorcism. If communicable spiritual essence is language, spiritual essence, at least in part, is not absent but undistinguishable from that which traditionally counts as the human domain *par excellence*. 'Linguistic essence *is* language' stands for the hope that spirit and essence will be present to those who are able to speak. In partaking of language, humans partake of essence and spirit; they live, as one is tempted to say, 'in truth'. By claiming that linguistic essence is immediately language, Benjamin tries to replace the scarcity of Kant's spiritual diet by a Hegelian cake, yet a cake baked without the bitter pill of dialectics. His preclusion of 'Vermittlung' ('mediation') agrees with both Hermann Cohen's dismissal of a mediatory *logos* and Edmund Husserl's assumption that essence is immediately present in intuition or *Wesensschau*.[18]

However, while the equation of linguistic essence with language is meant to bring essence closer to man, it actually removes 'language' from many instances of human communication. In telling me that lamb is 2.50 today, for example, my butcher communicates neither his language nor what is communicable of his spiritual essence. Of course, Benjamin does not maintain anything of the sort. He excludes the communication of facts from the notion of language by considering language the communication of *spiritual* contents, and he does not discuss, at least not explicitly, inter-human communication. He is concerned with the communication of nature (art, technology, etc.) to man and man's communication to God (140, 143–4; *62, 64–5*). However, to the same extent that he dislikes conversations about meat prices and the like,[19] my example is relevant: such dislike is likely to contribute to his decision not even to consider the extra-linguistic contents of language. On its own, of course, any dislike for what admittedly lacks in depth cannot lead one to claim that language or linguistic essence is the object of communication. This claim must draw its appearance of truth from experiences which seem to demand a description to the effect that in them language or linguistic essence is indeed communicated.

For any such description to be informative, Benjamin's initial introduction of linguistic essence does not suffice. As we saw, he introduces linguistic essence as that which is communicable of a spiritual essence. The description of an experience to the effect that something communicable is communicated, however, certainly threatens to fall under his verdict of tautological reasoning, no matter whether or not the something in question is spiritual. Let us therefore turn to that understanding of 'linguistic essence' that is implied in the following, strongly emphasised statement of the essay: '*There is no such thing as the content of language; as communication, language communicates a spiritual essence, i.e. a communicability as such*' (145–6; *66*). This explanation of spiritual essence also concerns 'linguistic essence', for the statement quoted is part of those reflections of the essay in which Benjamin comments on the hypothesis that spiritual and linguistic essence are identical. The notion of linguistic essence as communicability as such is compatible with his introduction of 'linguistic essence' if that which is communicable of a spiritual essence not only happens to be communicable but is so essentially. And even if everything communicable of a spiritual essence is essentially communicable, spiritual essence does not have to be communicable throughout. Now, as soon as we take 'linguistic essence' to denote communicability as such, we can think of experiences which can urge the experiencing person to maintain both that linguistic essence is communicated and that linguistic essence '*is*' immediately language. These are experiences where actual communication both seems unlikely and is of great emotional importance.

If a natural object, an artifact or a supernatural entity communicate, the very fact of that communication is highly significant for all those who – unlike Snow White's step-mother for example – do not take the possibility of such communication for granted. Its occurrence shows that the communicating entity is alive and that its life is, at least in part, open to human experience. In the context of everyday human communication, such disclosure is redundant and irrelevant. Enquiring about the price of lamb, I would rightly consider it a truism and out of place if my butcher made it the object of his reply that he is alive and that his life is not alien to mine. In the case of unlikely, non-human communication, however, the value of the corresponding 'information' may well outshine that of any particular content that may or might be communicated. In revealing the approachability of what otherwise may confront one as dead or alien, the experience of communicability in the non-human can be emotionally gratifying (as well as terrifying). Its value will appear the higher, the less self-confidence and pleasure one gains from one's communication with human beings. At the worst, the experiencing person can be so much in need of the gratifying reassurance that the experience of communicability creates that nothing else seems to matter. Affirming her desire or need, she may then claim that it is exclusively communicability as such which is communicated. If her desire and her attention are exclusively directed at communicability, moreover, the communicating entity's communicability will seem to be immediately given in that entity's actual communication. On the same condition, the actual occurrence of communication '*is*' the entity's communicability: it does not mean anything apart from the entity's communicability, its approachability or 'closeness'. Such closeness can be expressed by 'the entity communicates' as well as by 'the entity can communicate' or 'the entity is communicable'. For the experiencing person with her exclusive desire, any distinction such as that between possibility and actuality, faculty and performance, or *langue* and *parole* may therefore seem irrelevant. In the very first paragraph of 'On Language', Benjamin presents as identical two logically incompatible notions of 'language'. Language, he writes, 'means the principle directed at the communication of spiritual contents [...]. In one word: every communication of spiritual contents is language'. One word – and most probably any amount of words – is not enough to establish that something directed at x is x. Yet if the problematic approachability of an entity is the only thing that counts, the discovery of that entity's actual communication is as decisive as that of its orientation towards communication. It is the attitude of someone desperately eager for the closeness of spiritual essence which explains the fact that Benjamin neglects the logical difference. In so doing, however, he anticipates the equation of linguistic essence with language, or of communicability with communication, on which his notion of medium rests.

The experiences that provide Benjamin's notion of medium with a resemblance of truth are difficult to categorise. Although they agree with magical beliefs by implying the communication of entities which according to rationalistic assumptions do not communicate, they are not those of a magical consciousness, for 'communicability as such' will only be desired where communication is not taken for granted. Rather than speaking of magical experience we should therefore characterise what guides Benjamin as the, in Schiller's sense, 'sentimental' desire for an enchanted world. In their capacity for liberating the subject from the oppression of living in an alien and insipid world, the experiences in question furthermore resemble what, vaguely or misleadingly perhaps, has been termed 'aesthetic experience'.[20] However, they are neither an obvious response to all works of art or kinds of beauty, let alone ugliness, nor does Benjamin limit their impact to art or beauty. Mystic experience, too, provides support for Benjamin's notion of medium. It does so at least to the extent that Jürgen Habermas is right in holding that the mystic is more interested in his own feeling of God's closeness than in God himself.[21] Even if we allow for less egocentric mystics, it probably remains true to say that for the mystic God's communicability is valuable as such, it is valuable, that is to say, independently of what God may disclose about the world or *about* himself. On similar grounds, we may relate Benjamin's notion of medium to a particular case of inter-human communication. If I am in love with my butcher and if fulfilment is problematic, a Benjaminian philosopher who tells me that my beloved, in communicating, communicates his linguistic essence can easily win my ear. The philosopher, after all, would only confirm my inclination to take the beloved's communication, even if concerned with meat prices and the like, as a token of approachability. The claim that linguistic essence or communicability is the object of communication appropriately captures an erotic disposition like that, for instance, of Büchner's Lucile:

Camille: What do you say, Lucile?
Lucile: Nothing, I so much like to see you speak.
Camille: You hear me, too?
Lucile: But of course!

Camille: Am I right? Do you know at all what I said?
Lucile: No, truly, I don't.[22]

What the experiences corresponding to Benjamin's notion of medium have in common is that the approachability of their object is both uncertain and of great emotional value. In this, as well as in their aesthetic, religious and erotic dimensions, they agree with what he, in his mature work, calls 'aura'. This resemblance to his notion of aura notwithstanding, however, we probably do best not to categorise them as 'auratic experiences'. Benjamin's notion of aura is thoroughly ambivalent, and it is likely to be too rich, or too vague, to provide a reliable definition of one type of experience.[23]

Placed against the desire for communicability, Benjamin's contentions that linguistic essence is communicated and that linguistic essence *is* immediately language and vice versa lose their appearance of absurdity but remain misleading. They suggest that what is sometimes valued as the most important object of communication is communicated in actual fact, and is so always and exclusively. They conceal the subjective character of the underlying experiences, their restricted occurrence no less than the evaluation and desire at work in them, and, of course, the possibility of illusion.[24] Benjamin's notion of medium, that is to say, is not part of a metaphysics of language, nor can it lead to one. In 'On Language', the author's metaphysical ambitions characterise an attempt to present as real a world in which the desire for the presence of spirit is not only universal and exclusive but also always fulfilled.

1.2.3 'Name' and theology

In section 1.2.1 I argued that the equation of linguistic essence with language, which informs (L**), does not suffice to establish the truth of either (L*) or (L). As introduced by Benjamin, (L) presupposes the identity of spirit and language while the equation of linguistic essence with language does not exclude the non-identity of spirit and language. Yet the motive that leads Benjamin to state (L**) also explains why he, in stating (L), implicitly presupposes the identity of spiritual and linguistic essence. Whoever desires communicability of spirit is unlikely to be satisfied with the communicability of what may be only a part or one aspect of spirit. As long as spiritual and linguistic essence do not pass as identical, he may well suspect that in linguistic essence he reaches merely the appearance, and not the essence, of spirit. The desire to which Benjamin answers with his statement of (L), however, is no longer the unspecific desire for the approachability of any spirit but the religious desire for the presence of God or the absolute.

Let us start from Benjamin's observations on the infinity of language. This infinity, he maintains, derives from the immediacy of language. Because 'nothing is communicated *through* language, what is communicated *in* language cannot be externally limited or measured'. Benjamin concludes that therefore 'its own incommensurable unique infinity inheres in each language' and that the boundary of each language 'is defined by its linguistic essence, and not by its verbal contents' (143; *64*). This account rules out the relevance, to 'language', of an extra-linguistic reality such as that presupposed in 'bourgeois' theories. The verbal contents or extra-linguistic facts of 'bourgeois' theories, however, are not the only things that qualify for a non-linguistic reality. Anything incommunicable in spiritual essence also qualifies. If the spiritual essence that is communicated in language were in part incommunicable, the notion of language as the communication of spiritual essence would also refer to something that is not (throughout) language or communicability. Language would then be 'externally' limited by that aspect of spiritual essence which is incommunicable; it would be finite. In claiming language to be infinite, Benjamin implicitly rules out any difference between linguistic and spiritual essence. His characterisation of language as infinite can therefore be considered his commentary on (L).

This commentary shows a problematic displacement in the use of 'linguistic essence'. In contrast to his introduction of that term, Benjamin now considers 'linguistic essence' an attribute not of spiritual essence but of language. The boundary of language, we read, is defined by 'its linguistic essence', with 'its' referring to language.[25] To be sure, if both spiritual and linguistic essence, and linguistic essence and language, are identical, it may be formally correct to replace, while speaking of linguistic essence, the term 'spiritual essence' with 'language'. Yet it is precisely because of the preceding equation of language with linguistic essence that the reference to a linguistic essence of language is difficult to understand. In the light of that equation, the phrase 'its linguistic essence' comes to denote a linguistic essence of linguistic essence or a language of language. We are confronted with a problem of comprehension similar to that contained in the phrase 'Lucy's Lucy' if 'Lucy' is to refer to the same thing throughout.

We may not be able to understand reduplications of this kind, but we can describe what happens when Benjamin uses the notion of a linguistic essence of language after the equation of language with linguistic essence. What happens is that he credits language with that self-subsistence which traditionally characterises solely God or the absolute. Building on Aristotle as well as on neo-Platonic thought, Aquinas, for instance, considers God the self-subsisting existence ('esse ipsum subsistens').[26] This concept stands for the view that God, in contrast to any finite entity, which exists solely because it partakes of being, does not partake of being, but '*is*' being, and '*is*' being according to, or rather, '*in*', his own essence. God is that unique and incommensurable entity which through itself alone is determined in its essence, which essence, in turn, is both pure existence ('esse tantum') and pure plenitude. We may try to characterise this notion of God by saying with Cathérine Capelle that for Thomism in God, and in God alone, existence and essence coincide.[27] Or we may say that God's divine essence defines his (non-existent) boundary. The latter statement differs from Benjamin's notion of linguistic infinity only through the replacement of all references to language with references to God. Especially, the two components of the phrase 'God's divine essence' are meant to refer to the same thing, as do those of Benjamin's 'its [language's] linguistic essence' in the light of his equation of language with linguistic essence. The Thomistic God, after all, '*is*' his essence. The self-communication, moreover, with which Benjamin credits language is that God's essential quality.[28] The difficulty of comprehension with which Benjamin's characterisation of language as infinite confronts us is the familiar one of understanding the absolute. In this, too, his commentary on (L) agrees with (L). Language, if it communicates itself in itself, does not rest on anything other than itself, it is 'absolute' or 'infinite'.[29] In so far as (L) approaches the status of a definition,[30] language, according to (L), even is 'the absolute' or that which consists in its resting-on-itself. With his thesis of linguistic self-communication, then, Benjamin proposes a conception of the absolute, spelled out, however, not in terms of being but of communication. Accordingly, his notion of linguistic self-communication can be described as a divinisation of language. Using Kenneth Burke's phrase, we may say that 'language', in (L), is a 'god-term'.[31] The Epistemo-Critical Prologue to the *Trauerspiel* study supports this interpretation. Benjamin there uses the notion – held rather critically by his source Güntert – that Plato's theory of ideas rests on a deification of language to claim the linguistic character of the idea; in so doing, he simply disregards the difference between language and deified language.[32] Language, for him, is divine.

To establish as far as possible the divinisation of language occurring in

(L) is one of the major concerns in 'On Language'. This evaluation may look implausible, given the intricacy of the way that led us to recognise the God-like character of Benjamin's self-communicative language. What matters to me here, however, is not the esoteric dimension of Benjamin's presentation, but the fact that he is aware of the need to support what so far remained implicit and problematic in (L). Immediately after characterising language as infinite, he introduces a concept with which he tries to support the reduplication in his notion of the linguistic essence of language as well as his notion of medium and his presupposition of the identity of spiritual and linguistic essence. This concept is 'name'.

In line with Judaic thought, Benjamin sees in the bestowal of names a characteristic feature of human language.[33] We do not know, he maintains, of any non-human language in which entities are named (143; *64*). Thus introduced, 'name' seems important solely for human language. Since Benjamin will later argue that all differences between the various languages are merely differences of degree, however, 'name' is actually relevant to his general notion of language. As does (L), the concept of name concerns all language. Benjamin characterises its significance by stating that name is the quintessence or 'innermost essence' of language and something which can be called 'the language of language' (144–5; *65*).[34] Accordingly, Benjamin's reference in his commentary on (L) to the linguistic essence of language in the sense of 'the language of language' (or 'the linguistic essence of linguistic essence') is best understood as a reference to the quintessence of language. This means that the problematic reduplication on which his divinisation of language rests is as meaningful or nonsensical as is his concept of name. Benjamin presents 'name' as eminently meaningful, as the source, even, of meaning ('Bedeutsamkeit') itself.[35] 'Name', we may say, is surrounded by an aura of meaningfulness that extends, by virtue of the characterisation of name as 'the language of language', to the problematic reduplication in his commentary on (L).

If 'name', as used in 'On Language', refers to anything at all and if name has the significance which Benjamin attributes to it, then the reduplication on which (L)'s divinisation of language rests is, even if incomprehensible, justified. Yet Benjamin does not specify the sense in which he uses 'name' in the essay, nor does he disclose what 'name' refers to or why something of the sort should exist.[36] In his notes, he states conditions for passing as name which are so demanding that nothing may actually fulfil them: to pass as name, something must both inhere in the word and be a component of the corresponding object itself, a component, moreover, which is related to the object's essence (GS VI, 11–12; SW I, 88). As a result, the aura of meaningfulness that surrounds 'name' may well remain empty. The ways of language being what they are, however, this possibility does not obstruct the impression of meaningfulness. The 'God' of negative theology appears as eminently meaningful in spite, or because, of its stated lack of meaning.[37] Benjamin's auratic investment of 'name', at any rate, shows that he is aware of the circumstance that his earlier recourse to 'magic' does not suffice to support (L). Indeed, it takes more than a belief in magic to form a notion of the absolute. It takes theology.

The reader of 'On Language' is offered theology. (What she is not offered is clear evidence as to the author's beliefs about theology.) Relevant here are those theological assumptions that are related to Benjamin's use of 'name' in support of (L).

> Der Name ist dasjenige, [...] *in* dem die Sprache selbst und absolut sich mitteilt. Im Namen ist das geistige Wesen, das sich mitteilt, *die* Sprache. Wo das geistige Wesen in seiner Mitteilung die Sprache selbst in ihrer absoluten Ganzheit ist, da allein gibt es den Namen, und da gibt es den Namen allein. Der Name als Erbteil der Menschensprache verbürgt also, *daß die Sprache schlechthin* das geistige Wesen des Menschen ist; und nur darum ist das geistige Wesen des Menschen allein unter allen Geisteswesen restlos mitteilbar. [144; *65*]

> Name is that *in* which language itself communicates itself absolutely. The spiritual essence that communicates itself in name is language *as such*. Where the spiritual essence in its communication is language itself in its absolute wholeness, there alone is the name and there is the name alone. As the inheritance of human language, name therefore vouches for the fact *that language as such* is the spiritual essence of man; and solely for this reason is the spiritual essence of man the only one among all spiritual entities which is completely communicable.

According to (L), it is language in which language is communicated. According to the first sentence of the present passage, it is name. 'Name' therefore comes close to denote the medium language in that sense of the term (whatever it is) on which (L) rests. In the second sentence, Benjamin maintains explicitly the identity of spiritual essence and language (or linguistic essence), which he tacitly presupposed when stating (L). In the third sentence, name is said both to be the only thing existing in the realm of this identity and to exist nowhere else. As argued in section 1.2.1, (L) can be understood as resulting from the contraction of Benjamin's notion of medium with the equation of spiritual with linguistic essence. 'Name', in the passage under discussion, can therefore be considered the reiteration, condensed in one term, of Benjamin's postulate of linguistic self-communication.[38] With 'name', more precisely, Benjamin claims for human language what he claimed with (L) for 'every language' and what he will try to re-extend to all language later on in the essay.

If 'name' is to amount to more than a tacit repetition of (L), we need an answer to the questions why, in name, the self-communicating spiritual essence should be language, and why 'name' should be closely related to 'medium' in Benjamin's understanding of that term. The final sentence of the passage quoted last can be taken to contain Benjamin's answer to these questions. It contains Benjamin's answer if its phrase 'as the inheritance of human language' alludes to his Genesis commentary. Benjamin holds that the conception of language that he finds expressed in Genesis is true (147; *67*). According to his interpretation, Genesis conceives man's name language as originating in a transformation of God's creative word. It is this divine origin of human language to which the specification of name by 'the inheritance of human language' can be taken to refer. Name, accordingly, would be not (exclusively) a human heritage passed on from generation to generation but something which mankind originally received, or 'inherited', from God. Understood in this sense (and granted the truth of Genesis), 'name' is indeed, as Benjamin suggests, a reason for the identity, in man, of spiritual and linguistic essence.

According to Benjamin's Genesis commentary, the creative language emanating from God turns into knowledge in man (149; *68*). Knowledge is essential to man, for Benjamin calls man 'the knower' ('den Erkennenden'). Knowledge even counts as man's spiritual essence or, at least, as closely related to it. Otherwise the commentator could not suggest that the proposition that God created the knower in his likeness requires one to explain the equation of man's *spiritual essence* with language. The explanation that he actually offers is that man's spiritual essence is the language of creation. This contention, in 1916, is audacious. Benjamin after all presupposes with it that God created the world through the word. Taken in isolation, the contention is also heretical. Benjamin puts man in the place of God or man's spirit in the place of God's creative word. On the whole, however, the commentator pays tribute to theological correctness by holding that that which emanates changes its nature in the process, and by distinguishing accordingly between creation and knowledge, and between word and name. Man's spiritual essence is language, then, since it is the language of creation minus the latter's creative powers. The name language of man is God's creative language in a depotentialised state, or, as Benjamin puts it, human language is only the 'Reflex' of word in name.[39] This is why name, as the 'inheritance' of human language, 'vouches for' the identity of man's spiritual essence with language. And this is also why Benjamin can use the same verb 'to vouch' ('verbürgen') to characterise the relation between man's special connection with language on the one hand and the act of creation on the other. Using 'it' to refer to the act of creation, he states: 'it vouches for [...] the special connection between man and language from within the act of creation' (148; *68*). God's creation counts as the decisive and ultimate – the 'absolute', as it were – reason for the identity of man's spiritual essence with language, for the act of creation is said to vouch for that identity *from within itself*.

The Genesis commentary is the only part of 'On Language' from which it is possible to construct an argument in support of the equation of spiritual with linguistic essence (or language) that underlies (L). Except for his interpretation of Genesis and the related treatment of 'name' and human language, Benjamin's discussion of that equation is hypothetical in character (145–6; *66*). Although clearly sympathising with it, he does not argue for its truth but raises *the question* whether spiritual essence in general, and not only man's spiritual essence, is identical with linguistic essence. If we disregard two incidental occurrences of 'medium' (146, 157; *66, 74*), this hypothetical discussion of the equation of spiritual with linguistic essence is also the only passage outside his Bible exegesis where he returns to his problematic notion of medium, without, however, attempting an argument in its support. The support that his exegesis is apt to provide for (L), in turn, is clearly of a theological nature. Most probably, Benjamin's interpretation does not agree in all detail with any particular theological interpretation in a Judaic or Christian textbook. It does, however, agree with the doctrine of creation as such. It agrees with a doctrine, that is to say, which is not substantiated in 'On Language' independently from the Judaeo-Christian tradition of religious thought to which it belongs.

Benjamin's thesis of linguistic self-communication, in turn, is essential to 'On Language'. The requirements created by that postulate structure the development of thought in the essay to a high degree. And there is, apart from Benjamin's unelaborated claim that language is also the symbol of the incommunicable, no other single statement about language in this essay about language that equals (L) in range *and* importance. Since any argument which 'On Language' may suggest in support of (L) is theological, I conclude that the essay, *precisely if* read as a piece of theory, takes on a theological character. This means that if we are right to consider the essay a theoretical contribution – a contribution, that is, in which those assumptions are supported by argument that are both important and not obviously true – the essay is, and is essentially, theology.[40] This result conflicts to various degrees with the views held by critics who try to dissociate even this early text by Benjamin from the tradition of theological thought.[41] The result does not conflict with how Benjamin envisaged what he chose to call the coming philosophy. Since this alleged philosophy fulfils the demand for the oneness of philosophy and religion, theology may well be a better name for it, just as Benjamin conceded (GS II.1, 168; SW I, 108).

'On Language', however, also testifies to the author's unease about theology.[42] To take the theological character of the essay simply for granted can encourage an inadequate account of the problematic relation between Benjamin's writing and traditional religion. Benjamin is demonstrably interested in dissociating (L) from theological reasoning, even at the price of creating the appearance that (L) is a completely arbitrary postulate. Although he explicitly refers to scholasticism later on in the essay, the analogies between his notion of linguistic self-communication and the scholastic notion of God are nowhere mentioned. (L) is introduced without any reference to God or the doctrine of creation. Instead Benjamin attempts to have his divinisation of language pass as relying exclusively on 'immediacy' or 'the magic of language'. Dissatisfied with the result, for understandable reasons, he then turns to the concept of name. Historically speaking, there are not many concepts as highly charged with religious significance as this one. The equation of God's creative word with the name of God, for instance, which according to Scholem is essential to the cabbalistic notion of language,[43] provides a frame of reference for much of what is maintained in 'On Language'. Benjamin, however, does not discuss the name of God in the essay but instead introduces 'name' along apparently humanistic lines. In a note, in which he does discuss the name of God, he considers it to be 'no longer of a linguistic nature' (GS VII.2, 786). Before he reaches with his Genesis commentary the theme of the divine word, he declares the name language of man to be the only perfect language (145; *66*). And he does not begin his exegesis without the attempt to dissociate the argument of his essay from the traditional theological belief in the revelatory character of the Bible (147; *67*). The gap which is thus created between (L) and the theological assumptions that, if anything in 'On Language', support (L) is considerable. Specifications such as 'the purest sense', relating to Benjamin's understanding of 'medium', or 'the inheritance of human language', relating to name, can be taken to bridge that gap. Yet these phrases are inconspicuous and can of course be interpreted differently.

Benjamin's contention that human language is the only perfect one looks like, and probably is meant to be, a slap in the face of any respectable theologian.[44] The same holds for the pantheistic view, implied in (L), that *every* language is absolute. Traditional religious thought, both Judaic and Christian, must of course maintain the superiority of God over man, and of divine language over human language; and it must maintain the notion of God's irreducible otherness. Benjamin's provocations, however, amount to gestures that look as radical as they are powerless. He recants Promethean humanism by eventually declaring that the 'infinity' of all human language is of a 'restricted' kind compared to the absolute, creative and unrestricted infinitude of the divine word (149; *68*). Presumably, by the same token any non-human language that is not God's also ceases to count as absolute. As a result, 'On Language' oscillates between the extremes of the notion of God's otherness (or transcendence) on the one hand and pantheism and humanism on the other. In whatever way we shall eventually explain Benjamin's anti-theological gestures in the face of the theological substance of his thought, it is important that we do not misjudge that oscillation. It does not indicate the author's concern with the 'secularisation' of religious belief.[45] On the contrary, it is the re-enactment of a dilemma that is well known to traditional theology itself and which, for instance, informed the medieval concept of the analogy of being.

1.2.4 *Analogia linguae*

In stating (L), Benjamin tacitly presupposes the identity of spiritual and linguistic essence for all language. In his discussion of 'name', he maintains that identity explicitly, but only with regard to human language. He is aware of the fact that (L)'s general claim is still unsupported. Immediately after his treatment of human language he raises the question of whether all, and not only man's, spiritual essence is to be equated with linguistic essence. His answer is 'Yes – and No'. The differences between the languages, he maintains, 'are those of media which differ in their density, as it were, that is, in degree; and this with regard to the density of both the communicating (naming) and the communicable (name) in communication. These two spheres, neatly separated and yet united exclusively in the name language of man, of course correspond to one another constantly' (146; *66*). If all languages differ from one another only in degree, any (non-quantitative) predicate of one language can also, to some extent, truly be attributed to any other. It follows that if it is true to say of human language that spiritual and linguistic essence are identical, the same holds, to some extent, also for the other languages. 'Identity' can correctly be attributed in different degrees, or this is what Benjamin implies in his attempt to close the gap between the general identity claim, presupposed with (L), and his claim that in man spiritual and linguistic essence coincide.

In the present section I shall discuss, first, Benjamin's concept of density; second, I shall relate the implication that identity is quantifiable to that religious dilemma to which the medieval concept of *analogia entis* was meant to provide the solution; third, I shall include some historical observations on Benjamin's proximity to scholastic thought.

1. Benjamin's notion of density is closely related to what he, in Husserlian idiom, calls 'the essential law of language' ('das Wesensgesetz der Sprache'). According to this law, expressing oneself and addressing everything else are the same thing (145; *65*). Benjamin holds that the *Wesensgesetz* manifests itself in name since name is at the same time the ultimate exclamation ('der letzte Ausruf') and the true invocation ('der eigentliche Anruf') of language. It is possible that he relies in this on the idea that human essence finds its true self-expression in the invocation of God, in the invocation, that is, of the ultimate and all-comprising reality. Benjamin's law is also reminiscent of the notion that we express ourselves in the way in which we name things or, more generally, represent and confront the world in and with our language. The *Weltanschauung* that according to Humboldt is laid down in any particular language of course characterises those who speak it. Humboldt's influential idea, however, does not suffice to substantiate Benjamin's law. Even if we express ourselves by 'addressing' the world, expressing oneself and addressing everything else are not the same thing. The result is similar if we compare Benjamin's law with Friedrich Schlegel's idea that the novelist's intention to represent his time or the world around him can result in the novelist's self-representation.[46] External representation and self-representation are not the same even if an attempt at the former can result in the latter.

It is the so-called *Wesensgesetz der Sprache* that allows Benjamin to claim perfection for human language. In so doing, he for the first time refers to those two 'spheres' whose degrees of density will serve him as the criterion for distinguishing between the various languages. Intensive totality, he states, pertains to a language only if the spiritual essence that expresses itself in it is entirely communicable, that is, if its self-expression can exhaustively communicate what it is (145; *65–6*). And there is extensive totality, or universality, only if the essence in question names everything. Since according to the *Wesensgesetz* naming and self-expression are the same thing, the degree of intensity in a language and its degree of universality must also be the same. Therefore Benjamin can hold that a lack of communicability amounts to a lack of universality. Given that only the spiritual essence of man is entirely communicable, it follows that only man possesses a language that is perfect in both its intensity and universality.

The density of a medium, as we can now explain Benjamin's concept, consists in the degree of intensity and universality which it possesses. The more communicable a spiritual essence, the higher the degree of intensity which its language reaches. The more entities a spiritual essence names, the higher the degree of universality that characterises its language. The spheres and degrees of intensity and universality constantly correspond to each other because self-expression, the communication of what is communicable of a spiritual essence, and the bestowal of names to what is not the self-expressing entity are the same thing. Benjamin declares that this correspondence exists 'of course'. Yet it only exists 'of course' if we accept his *Wesensgesetz* to be self-evident. And only if that law is valid at all, does Benjamin's notion of a hierarchy of languages which differ in the degree of their two-dimensional density make sense. As argued earlier, that notion implies that there is to some extent identity of spiritual and linguistic essence in any non-human language. Therefore Benjamin can believe it apt to provide the missing link in the outlined substantiation of (L).

In fact, the missing link is not provided. Since Benjamin takes name to exist only where the spiritual essence in its communication is language or linguistic essence, the very assumption that non-human languages are situated in, or possess, a sphere of universality in the stated sense presupposes, but does not substantiate, that for them, too, spiritual and linguistic essence are identical. The second dimension of density, the sphere of intensity, is no more helpful. Given Benjamin's introduction of 'linguistic essence', the degree of intensity, being the degree to which a spiritual essence is communicable, is but the degree to which spiritual and linguistic essence are identical. Spiritual and linguistic essence, the author had maintained, 'are identical [...] *in so far* as the first is communic*able*'. Now, he merely renames such identity by calling it intensity. This may suggest an interpretation of the previous 'in so far' along quantitative lines; but it does not add anything in support of the assumption that linguistic and spiritual essence are indeed identical. What was needed to substantiate (L) was an argument for the identity of spiritual and linguistic essence in non-human languages; and what we get at best, even if we accept the validity of Benjamin's *Wesensgesetz,* is an argument which is unhelpful because of circularity or, indeed, a mere postulate of identity, and of a quantifiable identity besides.

2. The notion that there are various degrees of identity resembles that of a squared circle, for the sameness which the term 'identity' denotes excludes quantification.[47] Two entities, certain features of which are identical, are not themselves identical but similar. Yet the similarity between human language and what he considers non-human languages cannot satisfy Benjamin. If the expressive quality of natural, artistic, technological or historical phenomena only resembles human language, the term 'language' can be applied to them only metaphorically, and Benjamin, as we saw, categorically denies this qualification. Were human and non-human languages only similar, moreover, he would still have to argue that this similarity consists in the identity of the respective spiritual and linguistic essences. It is the identity of spiritual and linguistic essence in all languages which is needed to substantiate (L): an unspecified, or differently specified, similarity between the various languages will not do the job. Given that Benjamin offers no argument for the (straight) identity of spiritual and linguistic essence in non-human languages, it is appropriate to consider the idea of a quantifiable identity essential to his postulate of (L).

Although logically inconsistent, this idea is important in various philosophical paradigms, one of which is objective idealism. Benjamin, however, refers not to Hegel but to scholasticism. What he proposes is the notion of a hierarchy of languages similar to, and complementing, the scholastic conception of the hierarchy of being (146; *66*). With the latter conception, medieval philosophers responded to the conflicting ontological demands that resulted from their religious convictions. This can be illustrated by means of the notion of *analogia entis*.

On the basis of passages from Aristotle, scholastic philosophers discussed the question of how the sense ('ratio')[48] in which 'being' and other predicates were attributed to God and to finite entities respectively should be classified. Was this sense the same in both kinds of attribution or did the same word have different meanings when applied to finite as opposed to infinite entities? If the first was true, then the attribute in question was predicated *univoce*; if the second was true, then it was predicated *aequivoce*. For a believing philosopher, facing these options meant sailing between Scylla and Charybdis. The danger of univocation lay in the denial of God's irreducible otherness with regard to everything finite. Such a denial could lead to pantheism, or worse. Opting for irreducible equivocation, one ran the risk of denying that rational discourse about God was at all possible. This meant to play into the hands of agnostics or negative theologians. Any attempt to reduce the equivocation by distinguishing between the respective meanings in the predication of God and of things finite amounted to facing the danger of belittling God yet again. The meaning of the divine attribute, after all, had to be specified on the basis of human experience. The advocates of *analogia entis* proposed this idea to provide the way out. They held that 'being' was predicated of God and finite entities neither univocally nor equivocally but *analogice*. Thus Aquinas maintains that 'being' can be truly predicated of both God and any finite entity because both exist. Yet they exist in a different manner, and this difference affects the sense of 'being' in the two cases. While God *is* his being ('esse'), any finite entity exists only through participation in his being; it relies in its very existence on him. This difference precludes the univocal predication of 'being' in the two cases. Yet neither are we to assume straight equivocation, that is, two completely different senses of 'being'.[49] Rather one would maintain that the attributes of analogical predication are founded on *one* concept which realises itself, or is used, in various ways.[50] Since the various realisations or usages cannot be comprehended under any higher category, however, it is impossible to specify this concept. In the last analysis, it thus remains obscure whether or not analogical predications are founded on only one concept. It remains obscure, that is to say, how analogical predication can both, as Aquinas holds, be situated halfway between univocation and equivocation *and* use several senses of a word to signify different relations, even if these are different relations with regard to the same.[51]

The same objection holds here as with regard to Benjamin's implication of quantifiable identity. *Rationes* that are partly different are different; and *rationes* that are partly not different are similar and not the same, not even 'halfway' or 'not entirely'.[52] Yet similarity and, by extension, metaphorical predication can satisfy Aquinas as little as Benjamin. The two authors agree in their requirement that knowledge of matters divine be more than poetic. And they agree in their defence against the looming spectre of pantheism. Pantheism is not far off the road if, as Aquinas maintains, creation leads to what resembles the creator ('omne agens facit sibi simile'),[53] especially if this principle is linked to an argument for analogical predications which require more sameness, as it were, than the similarity on which metaphors are based. And pantheism is within reach when Benjamin locates God's creative word in everything existing (151; *70*). With (L), as we saw, pantheism has even arrived on the scene although, of course, as implicitly as Benjamin's linguistic absolute itself. Both authors, however, also fight the spectre which they risk summoning up. The Thomasian resort to similarity as opposed to identity – omne agens facit sibi *simile* – discards pantheism in the nick of time. And Benjamin does the same when he, eventually, accords (perfect) perfection only to the language of God.

3. Confronted with the main streams of Benjamin reception, my account of Benjamin's closeness to scholasticism may appear outlandish. I would therefore like to add two historical observations.

In a letter to Scholem, Benjamin characterises 'On Language' as immanently related to Judaism.[54] This characterisation of the essay is not invalidated by its explicit and substantial, if unelaborated, resort to scholastic metaphysics. Although Benjamin's reliance on the belief in analogical predication finds in Moses ben Maimon a very important Judaic opponent,[55] it is certainly true to say that scholasticism and Jewish religious thought share to a large extent both the medieval reception of Aristotle and the neo-Platonic tradition which inform the notion of *analogia entis*. Due to the genocide of Jewry in what once counted as Christian Europe, however, and due to Benjamin's related death, any reference to the common Judaeo-Christian tradition can appear obsolete. The exclusive label 'Jewish mysticism', in contrast, makes the religiosity of his writing more easily acceptable for many readers, especially Jewish and German ones, and those at odds with their respective religious orthodoxies.

The esoteric character of 'On Language' and the 'revolutionary' verve of the related 'Programme of the Coming Philosophy' can hide the fact that both essays, although not philosophical in form, agree with major concerns of the academic philosophy of their time. In the wake of neo-Kantianism, the extension of 'critical' epistemology to what Kant himself had regarded as matters not of knowledge but of faith was on the German academic agenda. And so was, in this context, the resort to scholasticism and its conception of *analogia entis*. 'The knowledge gained in mathematics and the natural sciences is not knowledge *tout court.'* Much of Benjamin's rejection of Kant's 'hollow' notion of experience reads like an echo of this statement in Heidegger's *Habilitationsschrift.*[56] Benjamin schedules both the Kantian and the Aristotelian table of categories for radical revision, Heidegger attempts to prepare the ground for the philosophical development of a so-called *cosmos* of categories which would overcome the metaphysical scarcity of such tables in general. In this attempt, he attributes central importance to the scholastic concept of *analogia*, on which Benjamin implicitly draws in 'On Language'. The latter's claim that knowledge should be epistemologically situated in a sphere of complete neutrality with regard to the subject–object opposition finds its Heideggerian counterpart in the allusion to a fundamental correlation between subject and object which would allow the philosophical appreciation of medieval mysticism. Before Heidegger's *Habilitation*, Lask had developed in considerable detail the outline of a logic that was to answer the categorical requirements of both transempirical speculation and empirical and scientific knowledge.[57] Although clearly less enthusiastic about the supernatural implications of medieval thought than Heidegger and Benjamin, Lask also relates the hierarchical structure of the envisaged logic to the Plotinian and scholastic ideas of *analogia*. Like Benjamin, both Lask and Heidegger are building on Husserl, and both are responding to the work of Benjamin's Freiburg teacher Heinrich Rickert. It is thus likely that Benjamin became familiar with their ideas before composing 'On Language'.[58] It is certain that he knew of those contemporary academic debates to which Lask and Heidegger contributed, given that he relates his own vision of 'the coming philosophy' to the common concerns of scholasticism and phenomenology. Placed against this background, Benjamin's hope that scholastic metaphysics would provide support for his thesis of linguistic self-communication is not in the least surprising.

1.3 Suspended revelation

1.3.1 Ambiguity as task

Since God is the highest spiritual essence (147; *67*), the best protection for his transcendence, which Benjamin implicitly denies with (L), lies in the incommunicability of spirit. In the penultimate paragraph of the essay, Benjamin introduces the notion of the non-communicable: in every case, language is 'not only communication of the communicable but also, at the same time, a symbol of the non-communicable.' The author does not elaborate on this new claim, especially, he does not discuss its problematic relation to (L). Nor do his notes clarify what 'a symbol of the non-communicable' can mean in connection with (L).[59] It is nevertheless reasonable to assume that by 'the non-communicable' he refers to something in spirit.

Expressions such as 'a symbol of the non-communicable' are traditionally used to indicate an indirect mode of presence that pertains to God or the absolute, especially after Kant relegated God from the realm of the knowable. Benjamin, too, uses the notion of what is incommunicable in the context of matters divine. In his letter to Buber of July 1916, that which is denied to the word coincides with 'the divine'. In the fragment 'On Riddle and Mystery' (1920–21), the symbol is conceived as consisting in, or resting upon, the acts of a living entity that is always God. In his notes, Benjamin unambiguously maintains that something spiritual exists which cannot be expressed. And the theological implications of *das Ausdruckslose* ('the expressionless'), a notion which he develops in 'Goethe's Elective Affinities' (1922), also support the assumption that the earlier 'non-communicable' concerns something spiritual.[60] Benjamin's eventual reference, in 'On Language', to what is non-communicable is therefore best understood as concerning an aspect or part of spiritual essence.[61] This interpretation implies that Benjamin implicitly revokes the equation of spiritual with linguistic essence that he tacitly presupposed when stating (L). If spiritual and linguistic essence are identical, there is in spiritual essence nothing non-communicable of which language could be the symbol. Conversely, if there is anything non-communicable in spiritual essence, (L), as introduced by Benjamin, cannot be true.

That Benjamin should eventually turn against (L) is understandable enough. With (L), as I have shown, he answers the desire for the presence of the absolute and even 'anticipates' its fulfilment. But this desire is unfulfillable in this world. Its unfulfillability at anything less than the price of death or apocalypse is well known, especially to those under its sway. The tragic hero, Benjamin writes in '*Trauerspiel* and Tragedy' (1916), dies 'since no one can live in the fullness of time' (GS II.1, 134; SW I, 56). No finite being can experience the absolute without losing – in the very process of experiencing – its absoluteness. To *experience* the absolute, if anything, is not to experience *the absolute*. To vanquish its absence, as Hegel puts it, is to suffer defeat.[62] With (L), the author of 'On Language' declares victory. With his notion of the non-communicable, he tries to mend the wreckage of his defeat.

Benjamin, however, presents the notion of language as a symbol of the non-communicable as a complement to his preceding discussion, and not as its partial revocation. If (L) is not to appear as *simply* obliterated at the end of the essay, he must not reveal that, in concluding, he has returned to the view dominant at its beginning, namely that spiritual and linguistic essence are not identical. It is therefore no accident that he does not elaborate on the notion that language symbolises the incommunicable. To avoid a corresponding discussion is even part of his strategy. The term 'strategy' is justified here, first because Benjamin's silence in the penultimate paragraph contributes to the creation of an ambiguous stance with regard to the equation of spiritual with linguistic essence which, apart from the discussion of 'name', characterises the entire essay; and second because the creation of that ambiguity is the author's avowed intention.

As to the first point, I briefly recapitulate. At an early stage of his essay, Benjamin warns against the equation of the subject of communication with its medium, that is, he warns against the equation of spiritual essence with language. He equally warns against the unqualified assumption of the identity of spiritual and linguistic essence. Since he will later maintain the identity of linguistic essence and language, both warnings tend to amount to the same thing. Furthermore, they are phrased in such a way that it remains possible both to maintain the identity of spiritual essence and language at a later stage of the discussion and to assume that of spiritual and linguistic essence in a qualified manner (Section 1.1). Benjamin then introduces (L). In so doing, he tacitly presupposes the identity of spiritual and linguistic essence for all languages (Section 1.2.1). As demonstrated by his subsequent treatment of 'name' and of the relation between the various languages, he is aware of this presupposition. His discussion of 'name' is the only place in the essay where he explicitly maintains the identity of spiritual and linguistic essence and his use of 'medium' shifts accordingly from implying the identity of medium and object to implying that of medium, object *and* subject. However, he restricts his explicit identity claim to human language (Section 1.2.3). The subsequent resort to scholastic metaphysics is to close the gap between the implicit and general assumption of the identity of spiritual and linguistic essence underlying (L), on the one hand, and the explicit claim of this identity exclusively for human language, on the other. If elaborated, Benjamin's reference to scholasticism amounts to the notion that there are different degrees of identity between spiritual and linguistic essence in the various languages (Section 1.2.4). Benjamin himself elaborates as little on his implicit notion of *analogia linguae* as he argues for his conception of name. On the basis of his Genesis commentary, an argument in support of 'name' and the identity of spiritual and linguistic essence can be constructed, but it is certainly not the author who offers this argument as such (Section 1.2.3). After the Genesis commentary (and the treatment of a number of related issues, which are immaterial here), Benjamin returns in the penultimate paragraph of the essay to his initial distinction between spiritual and linguistic essence. This return remains as implicit as did his general equation of spiritual with linguistic essence in the postulate of (L). It is confirmed, however, by the essay's concluding simile, according to which it is not language but a spiritual being, the sentry, who communicates his language in his language.

In the middle sections of the essay, then, Benjamin urges the acceptance of the general identity of spiritual and linguistic essence against which he had previously warned and which he will afterwards implicitly deny. Likewise in the middle of his essay, he maintains explicitly the identity of man's spiritual and linguistic essences. At no point of the essay, however, does he explicitly state, let alone argue for, the identity of spiritual with linguistic essence with the generality required to support (L). As a result, both the unrestricted equation of spiritual with linguistic essence and (L) hover in suspense. Benjamin, no doubt, postulates (L). And he clearly sympathises with the unrestricted equation underlying it. Yet whether he considers them true, we cannot tell.

This result is in perfect agreement with what Benjamin declares to be the task of linguistic theory. The following statement contains the programme of the essay itself:

> Die Ansicht, daß das geistige Wesen eines Dinges eben in seiner Sprache besteht – diese Ansicht als Hypothesis verstanden, ist der große Abgrund, dem alle Sprachtheorie zu verfallen droht, und über, gerade über ihm sich schwebend zu erhalten ist ihre Aufgabe. Die Unterscheidung zwischen dem geistigen Wesen und dem sprachlichen, in dem es mitteilt, ist die ursprünglichste in einer sprachtheoretischen Untersuchung, und es scheint dieser Unterschied so unzweifelhaft zu sein, daß vielmehr die oft behauptete Identität zwischen dem geistigen und sprachlichen Wesen eine tiefe und unbegreifliche Paradoxie bildet, deren Ausdruck man in dem Doppelsinn des Wortes Λόγος gefunden hat. Dennoch hat diese Paradoxie als Lösung ihre Stelle im Zentrum der Sprachtheorie, bleibt aber Paradoxie und da unlösbar, wo sie am Anfang steht. [141–2; *63*]

> The view that the spiritual essence of an entity consists precisely in its language – this view, taken as a hypothesis, is the great abyss by which all theory of language threatens to become captivated; to survive suspended right above this abyss is its task. The distinction between the spiritual essence and the linguistic essence in which the former communicates is the first and most basic distinction in any theoretical investigation of language. This distinction seems so unquestionable that it is rather the frequently asserted identity of spiritual and linguistic essence which forms a profound and incomprehensible paradox, the expression of which has been found in the double meaning of the word *Logos*. This paradox, as a solution, has nevertheless its place at the centre of linguistic theory, but remains a paradox and insoluble if placed at the beginning.

Hamann, one of the very few authors quoted in 'On Language', criticised his pupil Johann Gottfried Herder for exploiting the ambiguity of the word *logos*.[63] The teacher himself, however, did not refrain from putting that ambiguity to use. Benjamin's programmatic statement is reminiscent of the following passage in Hamann's letter to Herder of August 1784: 'Reason is language, *logos*. At this marrowbone I gnaw and shall gnaw myself to death. Even now, this depth lies in a persistent darkness for me; I still wait for an apocalyptic angel with a key to this abyss.'[64] Hamann's first sentence can be understood – especially if we take Benjamin's 'spiritual' to comprise, as in some idealistic authors, the meaning of 'mental' – as aiming at that identity of spiritual and linguistic essence with which Benjamin's programmatic statement is concerned. Hamann's imagery resembles Benjamin's. Both authors consider the equation of thought (reason or spirit) and language an abyss where destruction looms, but which nevertheless exercises an irresistible attraction. For both the equation is incomprehensible and both are positive that what they do not understand is profound. But there are also differences. While for Hamann the identity of reason and language is a truth which he does not hesitate to state, Benjamin requires linguistic theory to remain suspended 'right above' the abyss that is the equation of spiritual with linguistic essence. Theory ought to come close to stating the identity of spiritual and linguistic essence but it must not state it, at least not at the beginning. And while Hamann concedes, though not without self-irony, that he is waiting for supernatural and apocalyptic inspiration to provide the key to the riddle, for Benjamin the riddle itself is, somehow, the key. The paradox of the abysmal equation, we read, finds its proper place at the centre of linguistic theory, and does so, 'as a solution'.[65]

Benjamin's programmatic statement fulfils the task that he attributes to linguistic theory no less than the rest of his essay. Here, too, it remains undecided whether or not the general claim of the identity of spiritual and linguistic essence should count as true, and whether or not it ultimately remains a paradox. Benjamin calls the distinction between spiritual and linguistic essence 'ursprünglichst', that is, *the* first and most basic distinction, the distinction closest to an unspecified origin.[66] One might think that it is such partly because it is unquestionably true. Yet Benjamin maintains that the difference between spiritual and linguistic essence only 'seems' to be unquestionable. It seems so much so to him, however, that he considers its opposite, the identity of spiritual and linguistic essence, an actual paradox. It 'forms', and does not only appear to form, a paradox. In claiming that 'this *paradox*' (my italics) serves as a solution at the centre of linguistic theory, he suggests that what at the beginning is a paradox also remains one 'at the centre'. Yet through the subsequent 'but' in 'but remains a paradox and insoluble if placed at the beginning', he superimposes the solution of the paradox to the solving paradox of the preceding clause.

Without doubt, in 'On Language' the idea of name is the 'centre' where the paradox of the identity of spiritual and linguistic essence finds its place. Only where spiritual and linguistic essence are the same does name, the innermost essence of language, exist; and only with regard to the name language of man does Benjamin explicitly maintain the identity of spiritual and linguistic essence. His programmatic statement, however, requires not only that the paradox of that identity be situated at the centre of theory but also that it be solved there. As the place of this solution, the 'centre' of linguistic theory does not lie within 'On Language'. As far as the author is concerned, the solution of the central paradox, the acknowledgment of which he programmatically requires, is not part of his job.

1.3.2 Revelation

Why is it that theorists of language should sacrifice the possibility of obtaining results to the sophisticated, yet frustrating exercise of Benjaminian ambiguity? This is a rationalist's question. Why make so much of Genesis if there are reasons to believe in the corresponding view of language which are not based on the assumption of biblical revelation? With such a question a rationalistic theologian – a believer, that is, in natural theology – may respond to 'On

Language'. And a believer in the indispensability of biblical revelation may ask why we should discard with Benjamin the belief in the Bible's revelatory character if we hold, equally with Benjamin, that the view of language contained in it is both true and inexplicable. Such questions find their answer in Benjamin's attitude towards the concept of revelation. In the present section, I shall examine the expression of that attitude in 'On Language' in four steps. 1. I shall discuss Benjamin's claim that the equation of spiritual with linguistic essence leads to the concept of revelation and characterise it as based on an aesthete's desire for the religious re-enchantment of modernity. 2. I shall analyse Benjamin's introduction of his Genesis commentary; this introduction demonstrates three things, namely the logical presupposition that the belief in biblical revelation is true, Benjamin's actual disbelief, and his hope for divine revelation. 3. I shall explain the programmatic prescription of ambiguity and the related postulate of linguistic self-communication on the basis of Benjamin's attitude towards 'revelation'. 4. I shall conclude the present section by formulating an alternative to Benjamin's view of the relationship between the equation of spiritual with linguistic essence and the concept of revelation, and by some remarks on the historical context of his concern with revelation.

1. Benjamin's considers the equation of spiritual with linguistic essence of far-reaching metaphysical consequences because 'it leads to that concept which again and again, as if of its own accord, has arisen at the centre of linguistic philosophy, and which has provided the most intimate connection between the philosophy of language and that of religion. It is the concept of revelation' (146; *66*). The historical implications of this claim are unconvincing. The concept of revelation is irrelevant to much of what, in 1916, can reasonably count as philosophy of language. More importantly, 'revelation' never arose 'as if of its own accord' in the mystico-theological tradition on which Benjamin draws. A claim such as Hamann's 'reason is language, *logos*' can only lead to the concept of revelation where the author presupposes a religious view of language and reason in the first place. The mere assumption of an identity of mental and linguistic processes, in contrast, is devoid of that religious significance which characterises 'revelation'.

Benjamin's deduction of the concept of revelation from the equation of spiritual with linguistic essence comprises three steps:

1. The equation implies that 'the deeper, i. e. the more existent and real spirit is, the more expressible and expressed it is, too'.
2. It follows from this implication, that 'the linguistically most existent, that is the most fixed, expression, that which is linguistically most concise and unchangeable, in a word, that which is most expressed is at the same time the purely spiritual'.
3. This consequence is what 'the concept of revelation means when it takes the inviolability of the word to be the sole and sufficient condition and characterisation of the divinity of that spiritual essence which expresses itself in it'. (146–7; *66–7*)

Benjamin's unusual comparatives and superlatives of 'existent', 'real', 'expressible' and 'expressed' result from the envisaged extension of the scholastic 'hierarchy of being' to linguistic theory. The equation of spiritual with linguistic essence, he holds, implies that spirituality and expressibility are directly proportional. The notion which is thus discarded is that of an *inverse* proportionality between spirituality and expressibility. If there is identity of spiritual and linguistic essence, Benjamin argues, it is not the case that the more spiritual or 'deeper' something, the less expressible it is. Spirituality being nothing but communicability, the more communicable something is, the more spiritual it is, too. This implication of the equation of spiritual with linguistic essence, according to Benjamin, is the meaning of the term 'revelation'.

Benjamin does not relate his understanding of 'revelation' to any particular theological doctrine. This may partly be due to a lack of knowledge,[67] but there is also a substantial reason. His comment on the meaning of 'revelation' is strikingly irreligious. For anyone who believes in God's self-revelation in the Scriptures, the inviolability of the biblical word is certainly not a condition of God's divinity, let alone the sole and sufficient one. Rather it is God's divinity that makes the revelatory word inviolable. Nor does it testify to any religious understanding if the maximum of linguistic precision or conciseness comes close to being taken as the reason for believing in the inviolability of the biblical text. Precision is an aesthetic criterion, and to experience an expression as perfect and unchangeable is aesthetically rewarding. The believer in divine revelation, however, does not take the word of Scripture to be inviolable because God is such a good stylist.[68] Benjamin's appeal to conciseness echoes the aestheticism in the circle around Stefan George, and so does his use of the term *Gestaltung* ('formation') in the present context.[69] His deduction, then, actually does not lead to the concept of revelation. It leads at best to only one implication of 'revelation', namely the impression of unchangeability that the religious believer in the revelatory word shares with the linguistic aesthete who admires, say, the perfection of a poem.

In spite of his reliance on aesthetic features, however, Benjamin does not consider the religious significance of 'revelation' as negligible in linguistic theory. Otherwise he could not call the connection between the philosophies of language and religion, which he locates in the concept of revelation, 'most intimate'. Nor could he, by an affirmative quotation from Hamann, indirectly draw on that text which, being closely related to Hellenistic Judaism, became canonical for a Christian understanding of language and revelation, the preamble to St John's Gospel.[70] With regard to 'revelation', then, the essay displays two features of the author's attitude, the simultaneous occurrence of which we need to understand. Benjamin explains the meaning of the term 'revelation' in an irreligious, namely aestheticist, manner, and he wants to retain its religious significance.

The first feature means that we cannot think of Benjamin as an actual believer in divine revelation. The second feature rules out the view that Benjamin intends to secularise religious thought. General misgivings about the notion of secularisation in intellectual history put aside,[71] this view could in principle explain Benjamin's aestheticist account of 'revelation' and related characteristics of 'On Language'. Benjamin's secularisation would consist in using 'revelation' for the aims of aesthetic theory or poetics. Thus Menninghaus, for example, holds that Benjamin's secularisation of the mystical tradition of linguistic thought leads to a central distinction between the referential and poetic (or self-referential) functions of language, which is 'largely equivalent' to that found in Russian Formalism and (post)structuralism.[72] In actual fact, however, Benjamin explicitly rejects the idea that it is in art and poetry – and not exclusively in religion – that that perfect identity of spirit and language can be found which underlies his notion of linguistic self-communication.[73] In its notion of revelation, he argues, religion is the only realm that does not know the inexpressible; only God's spirit is purely linguistical, while the spirit expressed in art and poetry is necessarily tainted by its communion with matter (147; *67*). Of course, Benjamin's reasoning here is based on shaky ground. The idea that God reveals himself *as the ineffable* is prominent in the religious discourse on revelation. Yet Benjamin also pays tribute to that idea by maintaining that it is *the inexpressible* which expresses itself as revelation.

Textual evidence thus rules out the assumption that 'On Language' is informed by an attempted secularisation of religious thought *and* it rules out the assumption that the essay is based on the author's actual belief in divine revelation. I therefore propose to explain the coexistence of areligious aestheticism with Benjamin's theoretical affirmation of religion in terms of hope. The aesthete Benjamin hopes for God's future revelation or a future belief in divine revelation. Therefore he emphasises the importance of the religious idea of revelation in his essay on language. Being an aesthete without actual belief, however, he cannot help explaining the meaning of 'revelation', of all things, in a non-religious and aestheticist manner. While secularisation, in Hegel's words, tries to regain for the earth what was sold at a loss to heaven, the hope for God's (exclusively) future revelation rests on a desire for the sacred in what is actually experienced as a secular world.

2. Benjamin's writings before 1916, which I shall consider in some detail in sections 1.3.3 and 2.1, provide ample and clear evidence for the assumption that his early thought is informed by a desire for the religious re-enchantment of modernity. In 'On Language', positive textual evidence for my interpretation is more difficult to unearth. However, it is contained in Benjamin's introductory remarks to his Genesis commentary.

> Wenn im folgenden das Wesen der Sprache auf Grund der ersten Genesiskapitel betrachtet wird, so soll damit weder Bibelinterpretation als Zweck verfolgt noch auch die Bibel an dieser Stelle objektiv als geoffenbarte Wahrheit dem Nachdenken zugrunde gelegt werden, sondern das, was aus dem Bibeltext in Ansehung der Natur der Sprache selbst sich ergibt, soll aufgefunden werden; und die Bibel ist *zunächst* in dieser Absicht nur darum unersetzlich, weil diese Ausführungen im Prinzipiellen ihr darin folgen, daß in ihnen die Sprache als eine letzte, nur in ihrer Entfaltung zu betrachtende, unerklärliche und mystische Wirklichkeit vorausgesetzt wird. Die Bibel, indem sie sich selbst als Offenbarung betrachtet, muß notwendig die sprachlichen Grundtatsachen entwickeln. [147; *67*]

> If in what follows the nature of language is studied on the basis of the first chapters of Genesis, it is neither intended to pursue the purpose of exegesis nor to base the discussion in the present essay objectively on the Bible as revealed truth, but to discover what emerges from the biblical text with regard to the nature of language; and *for the time being* the Bible is only indispensable in this because the present

> comments follow it in the basic idea of presupposing language to be an ultimate, inexplicable and mystical reality, perceivable only in its unfolding. The Bible, in considering itself revelation, must of necessity develop the fundamental facts of language.

It is clear from this that Benjamin tries to exempt the argument of his essay from the belief in biblical revelation, which, since the Enlightenment, formed a main target of the critique of religion. It is not, he maintains, that his argument is based on the belief in the revelatory character of the Bible: the Bible is only indispensable for his essay because he follows it in presupposing language to be a mystical reality. This justification sounds as if Benjamin did not wish to draw on the Bible but was forced to propose a Bible-based approach to language which, being such, unfortunately happened to be in line with the belief in biblical revelation. What should coerce him in this way? The Bible, he claims, is indispensable. Yet the reason that he gives in support of this claim is obviously invalid.

The fact that Benjamin agrees with the Bible in presupposing language to be a mystical reality of course does not make the Bible indispensable. One can axiomatically assume a mystical view of language, one can even share the specific mystical presuppositions contained in the Bible without actually drawing on it. To be sure, the Bible is indispensable for the discovery of what emerges 'from the biblical text' about language and Benjamin refers to such a discovery in his argument by the phrase 'in this' ('in dieser Absicht'). Yet he had stated just before that exegesis is not his aim. And surely, his essay is not a doxographical work on the biblical authors but attempts a philosophical outline concerning the nature of language. Any philological indispensability of the Bible is therefore irrelevant.

The last sentence of the passage quoted above may also be taken as aiming at support for the alleged indispensability of the Bible. If the Bible, in considering itself revelation, must of necessity develop the truth about language, its interpretation is a philosophically fruitful method to understand the nature of language. Benjamin may thus attempt to bridge the gap between philosophical and philological requirements, the neglect of which gap is in part responsible for the weakness of his preceding reasoning.[74] Yet any such attempt at bridge-building would be insufficient to substantiate the claim of the Bible's indispensability. Let us concede that the Bible, in considering itself revelation, must of necessity develop the facts of language. It does not follow from this that the Bible is indispensable for any such development to take place. On the same grounds the latter may be said to occur in any text that claims to be (based on) revelation, no matter whether it is handed down to or composed by the (mystic) philosopher. It is noteworthy that Benjamin, in the very attempt to argue for the indispensability of the Bible, actually argues for the truth of *any* allegedly sacred book or text. This should prevent us from seeing in his views on, and use of, the Bible a religious commitment to Judaism or Christianity.[75]

The last sentence of the passage is insufficient to substantiate Benjamin's indispensability claim. With regard to his attempt at exempting his essay from the rationalistic criticism of the belief in biblical revelation, however, the sentence is fatal. It presupposes the truth of that belief. Benjamin infers from the biblical belief in biblical revelation that the Bible of necessity unfolds the truth about language. This conclusion is only correct if the biblical belief in biblical revelation is true. If it were false, both the true (the 'facts') and the false (the errors about language) could follow (or be 'developed') from it; the truth about language would not follow of necessity. If, in contrast, the biblical belief in biblical revelation is true, whatever actually follows (or is correctly 'developed') from it is also true. Hence it is, as Benjamin claims, necessarily true. It is Benjamin's reference to necessity which is responsible for the implicit presupposition that the belief in biblical revelation is true. I do not think that he would be prepared to drop it, given the insistence with which he appeals to necessity twice over: 'must of necessity'. He cannot drop it easily. If we omit 'must of necessity', his sentence reads: The Bible, in considering itself revelation, develops the fundamental facts of language. The question immediately arises why this should be so. It is unlikely that it could be answered without doing what Benjamin's avoids doing throughout, namely revealing frankly whether or not he actually believes in what he implicitly presupposes: divine revelation.

One may object to the preceding argument as follows: Benjamin does not formulate his point to the effect that true propositions about language follow from the biblical assertion that the Bible is God's revelation but refers to a development ('Entwickeln') of the facts of language from within the biblical view of the Bible. He uses not the frozen metaphor 'Entwicklung', as it denotes the logical evolution of thought or argument without intended or essential reference to life, but a mystical notion of 'Entwicklung', which corresponds to the 'unfolding' ('Entfaltung') of his previous sentence in that the reference to life is essential.[76] The argument applies the rules of logical conclusion to a sentence that is not written in the language of logic. It is utterly futile to argue against the truth of religion on the basis of the logical incorrectness of religious discourse, futile for both the believer in divine revelation and the believer in logic.

The present case, however, is different. Benjamin does not use the language of logic, but neither does he use the language of faith. The argument, moreover, is not concerned with the truth or falsehood of the belief in revelation. It uses the rules of syllogism as a heuristic device to show that Benjamin, in the very process of denying it, presupposes the belief in biblical revelation to be true. Still, it is desirable to explain the sentence in question on the basis of Benjamin's own wording and with particular reference to its subject matter, the Bible. This can be done.

> IM ANFANG WAR DAS WORT/ UND DAS WORT WAR BEY GOTT/ UND GOTT WAR DAS WORT. ... Alle ding sind durch dasselbige gemacht [...]. In jm war das Leben/ und das Leben war das Liecht der Menschen [...]. (John 1: 1–4)
>
> In the beginning was the Word, and the Word was with God, and the Word was God. [...] All things were made by him [...]. In him was life; and the life was the light of men.

These verses agree with Benjamin's mystical implications in presenting knowledge ('light') as a reality. Since the Bible, as Benjamin maintains, considers itself a revelation, John 1: 1–4 can be read as characterising the words of the Bible. An interpretation of the quoted verses as self-referential, especially on the basis of Luther's translation, yields that in John's preamble the, in Benjamin's words, 'facts of language' are considered to be contained in, and developed from, the words of the Bible.[77] Benjamin, however, goes further. He does not claim that the Bible presents the facts of language as inherent in the biblical word; he claims that the facts of language inhere in it. This, again, he can only do if he presupposes the biblical belief in biblical revelation to be true.

So far, I have disregarded the emphasised qualification *'for the time being'* (*'zunächst'*) in Benjamin's introduction of his Bible commentary. It is not only the emphasis itself, but also the use of the same phrase in his earlier (and then suspended) warning against the equation of spiritual essence with language that gives us every reason to consider it twice. Because of the emphasis, the German sentence in which it occurs can easily be misunderstood as meaning that the Bible is indispensable only initially.[78] Yet what it actually says is that Benjamin's explicitly stated reason for the indispensability of the Bible is only 'for the time being' the only one. In considering the alleged reason for that indispensability as only temporarily exclusive, Benjamin envisages a different reason for it in the future. At no point in the essay does he spell out this future reason, but we are nevertheless entitled to take it to be the valid one. The Bible is indispensable for an investigation into the nature of language only if it is the revelation of a truth which cannot be obtained by any non-biblical means, for instance by the human understanding at work in philosophical and linguistic enquiries. The Bible, that is, is indispensable only if it contains, is, or is based on, superhuman or divine revelation. We are entitled to see in the future reason which Benjamin envisages the truth of the belief in biblical revelation mainly for two reasons. First, he considers language an inexplicable reality, developed *as such* in the Bible. And second, as I have argued above, he implicitly presupposes the truth of the belief in biblical revelation even where he is explicitly engaged in discarding it.

Of course, the attempt to accommodate the rationalistic criticism of the belief in revelation betrays an actual scepticism on Benjamin's part as regards the truth of that belief; and so does his areligious comment on the meaning of 'revelation'. Now, no sceptic can rule out the possibility that one day he will believe in revelation. Only the sceptic who desires to do so, however, will pay tribute to this possibility in his writing. Therefore Benjamin's attitude towards the idea of divine revelation is best characterised as combining actual disbelief with the desire to believe. He desires God's future revelation without believing in a revelation which, according to Judaic and Christian beliefs, already occurred in the transmission to humanity of the Torah or the Bible (or in some of the allegedly historical events related in them). If pressed on this point, however, the author of 'On Language' would have to concede that even the traditional beliefs in the sacredness of these particular texts may be justified by the advent of that revelation which he desires.

3. For Benjamin, the equation of spiritual with linguistic essence leads to the concept of revelation. We can therefore draw on his attitude towards the idea of revelation in order to explain his prescription, to linguistic theory in general, of an ambiguous treatment of the equation of spiritual with linguistic essence. Whoever is attracted by, and shrinks from, the fire will also be attracted by, and shrink from, what he rightly or wrongly considers the way leading to it. If, moreover, he is self-

confident or hesitant enough to consider his own ambivalence an expression of truth, he will tend to impose it on others.

All linguistic theory, according to Benjamin, must hover *suspended* above the equation of spiritual and linguistic essence because the unequivocal assertion of their identity leads to the notion of divine self-revelation, a notion which disagrees with both Benjamin's own disbelief and the general religious scepticism which characterises European intellectual culture after the Enlightenment. By unequivocally asserting that identity, the philosopher of language would thus compromise his intellectual integrity. He would also compromise the idea of revelation itself, for the identity of spiritual with linguistic essence can truly be asserted only if revelation already occurred. In asserting that identity, then, he would accord revelatory status to events or texts which, for Benjamin, do not deserve it.

All linguistic theory must hover suspended *right above* the equation of spiritual and linguistic essence because this equation leads to the religious concept of revelation which corresponds to the object of Benjamin's desire. As we saw, it is only in the religious experience of divine revelation that he considers spirit to reveal itself untainted by any inexpressibility. Only this experience allows man to partake of the unrestricted presence of pure spirit. Or so Benjamin, the sceptic desiring belief, assumes.

The equation of spiritual with linguistic essence means that spirituality gains in terms of accessibility. By the same token, however, spirituality loses in terms of an irreducibility to human understanding. For all his struggles for 'the crystal clear elimination of the unsayable' (GB I, 326), Benjamin deeply cherishes the promise of depth that is traditionally linked to the notions of both the spiritual and the inexpressible. He praises language as an inexplicable, mystical reality. And he creates an aura of meaningfulness around 'name' without specifying the sense in which he uses the term. With depth and its secrets gone, the equation in question safeguards only the accessibility of a spirit the revelation of which looks futile. Therefore, too, linguistic theory must leave that equation suspended.[79]

With his prescription of an ambiguous stance with regard to the equation of spiritual with linguistic essence, then, Benjamin urges what he calls linguistic theory to pay tribute to the central paradox in the notion of revelation, namely that God's inexpressible spiritual essence expresses itself. If we concede that theory is a field where religious aspirations can be expressed, this is understandable enough. For the paradox of revelation cannot be solved by theoretical means. If it exists at all, it can only cease to exist together with God's transcendence, his irreducible otherness, or the finite constitution of this world. Benjamin's 'theory of language', then, is eschatological in character. As in Hamann, the equation of spirit with language is considered an abyss since – for Benjamin, too – it is pregnant with apocalypse. With his postulate of linguistic self-communication, the writer even 'anticipates' the end of the world; for as regards (L), he does not take the same precautions as with regard to the underlying equation of spiritual with linguistic essence. While (L), in the context of the essay as a whole, partakes of Benjamin's eschatological suspense, he does affirm linguistic self-communication in positive, unambiguous terms on the occasion of its introduction. For readers who are unhappy with the splendours and miseries of phantasmagoric wish-fulfilment, this makes (L) a forced and misleading expression of his desire. Yet a reader who acknowledges the author's assumption that man, in human language, communicates his spirit to God can see in the writing of (L) – and, to a great extent, in the writing of the essay as a whole – an expression of the desire for revelation *that is addressed to God*. For Benjamin, at any rate, the writing of 'On Language' must have resembled a prayer for divine revelation, for in the dissociation of its 'immediate' prose from the purposes of worldly action as well as in its attempt at sanctifying the profane, the essay agrees with that view of prayer which he affirms in his letter to Scholem of 15 September 1919. Citing a cabbalistic source from Ernst Bloch's *Spirit of Utopia*, Benjamin writes:

> 'Know that a double view exists for all worlds. One shows their external side, that is, the general laws of the worlds, according to their external form. The other shows the inner essence of the worlds, that is, the quintessence of the human souls. There are therefore also two degrees of action, the works and the orders of prayer. The works exist to make the worlds perfect with regard to their external form; the prayers, however, exist to include the one world in the other and to elevate it.' *Never have I read anything plausible about prayer, apart from this.* [GB II, 45, my italics]

4. With (L) and its invocation of the absolute, Benjamin draws on traditional concepts of revelation while expressing at the same time a desire which, for him, finds no fulfilment in traditional religions. According to Scholem's reconstruction of cabbalistic belief, it is the mystic Torah or God's name in which God reveals himself. God, furthermore, is in his name and his name is in him, he is his name.[80]

Scholem's cabbalistic God, that is to say, reveals himself in himself just as Benjamin's language communicates itself in itself. And a Christian, supported by the dogma of God's triune nature, believes that in revealing himself in Christ, the *logos*, God reveals himself in himself. With (L), Benjamin ascribes to language a structure which pertains to the idea of divine revelation in the Judaeo-Christian tradition, an idea, which this tradition itself – especially in its neo-Platonic dimensions – already expresses in a linguistic vocabulary. The laboured identities, in 'On Language', between the subject and the object of communication, and between both and the medium of communication possess their historically powerful predecessors in traditional belief.[81] While Benjamin is wrong, then, in suggesting that the mere philosophical equation of spiritual with linguistic essence leads to the concept of revelation, there exists indeed a close connection between the traditional belief in revelation and that equation. At least where the notion of God's ultimate and unrestricted self-disclosure is concerned, a logical analysis of the (alogical) belief shows it to imply the identity of spiritual and linguistic essence. We may therefore invert Benjamin's claim and say that the 'concept' of revelation leads, by way of analysis, to the equation of spiritual with linguistic essence.

In its concern with divine self-revelation, 'On Language' agrees with a characteristic feature of modern theology. At least since the eighteenth century, theological thought is characterised by the increasing emphasis which it puts on the notion of divine self-disclosure at the expense of an understanding of 'revelation' according to which revelation consists in a number of propositional truths concerning various subjects such as God, nature or the rules of ritual, moral or social life.[82] This development, which would reach considerable proportions in the twentieth century, was to some extent prepared by Jacob Böhme's notion of divine self-revelation,[83] his influence on the Swabian Pietists, and their explicit reliance on the cabbalistic doctrine of the ten *sefirot*.[84] It includes Schelling's notion of 'communicativum sui' as the principle of revelation and his equation of the revealing subject with the revealed object no less than Franz von Baader's critical defence of that equation.[85] In the nineteenth century, Richard Rothe claims that God is the sole object of divine revelation.[86] For Isaak August Dorner, God's 'perfect self-communication' has even become essential to a *dogmatically* correct understanding of God's nature.[87] And in the twentieth century the notions of divine 'self-communication' or 'self-revelation' are of central importance for instance in the works of Max Scheler, Paul Tillich, Rudolf Bultmann, Karl Barth and Karl Rahner.[88]

The fact that modern theology is so much concerned with the rationalisation of the traditional belief in divine self-revelation is understandable enough. The more religious scepticism increases in a society, the more the advocate of God's existence has to count on the latter's self-disclosure. The competing options have little appeal. The belief in cosmogenic revelations faces in modern science so strong an opponent that it is wise to avoid direct confrontation. Revealed truths *about* God will find possible acceptance only to the extent that God's very existence is accepted. And as regards the traditional belief in divinely revealed laws, its rationalisation also depends on the belief in divine self-revelation. A possibly non-existent entity has poor prospects of passing as a lawgiver.[89] If modern life is to become religious again, the existence of what can sanctify it needs to be revealed in the first place. Perhaps Franz Rosenzweig's puzzling notion that God's self-revelation forms the object of a necessary second order revelation can be explained along similar lines.[90] Benjamin's concern with divine self-communication certainly can, for in Benjamin it is the actual sceptic who desires his life to be sanctified.

1.3.3 The sceptic's prayers

As a prayer for revelation, 'On Language' amounts to the esoteric condensation of that religious rhetoric which characterises Benjamin's earlier writing in the context of the German Youth Movement.[91] The manifesto 'The Religious Position of the New Youth', published in *Die Tat* ('The Deed') in 1914,[92] demonstrates both his determination to lead a holy life and his actual experience of Wilhelmine society as secular. '*A generation desires to be again at the parting of the ways, but the bifurcation is nowhere to be found.* [...] The new youth faces a chaos in which the objects of its choice (the sacred ones) are disappearing. No "pure" and "impure", "sacred" and "depraved" lights its way, only the schoolmasterly words "permitted-forbidden".' Youth 'demands nothing more urgently than the choice, the possibility of choice, the sacred decision as such' (GS II.1, 73). Along similar lines, Benjamin had already affirmed the thirst after a religious and cultic exaltation of life in his dialogue on religiosity: 'Religion provides a new foundation and a new ennoblement for daily life, for convention. It turns into cult. Are we not thirsting after a spiritual, cultic convention?' (GS II.1, 29).

Thirsting after the cultic exaltation of the everyday, Benjamin's 'we' long for a service that traditional religions and their Wilhelmine heirs try to provide. Yet he rejects 'the old religions', Judaism no less than Christianity, as socially opportunistic and not truly religious. On the same grounds he pours ridicule on the Congress of Monists and its intention to provide speakers for monistic edification at family celebrations.[93] By 1912 at the latest, he had refused all the offers of a more than momentary spiritual fulfilment which contemporary culture held in stock. The beliefs, inherited from the Enlightenment, in egalitarian humanism and moral autonomy, in progress or, at least, evolution, fare no better in his judgement than the traditional religions, monism or what to him is the mawkish pantheism of the Wilhelmine establishment: they all fail to guarantee 'something eternal in our daily work' (GS II.1, 17, 20, 22, 25). Either they do not draw a dividing line between the sacred and the secular such as prevailed in pre-modern society or they do not meet, in their attempts at drawing such a line, the intellectual standards of what Benjamin suggests is the cultural elite of his time. His 'geistige Führer' ('spiritual leaders')[94] share the fate of the first of the three youths in his 'The Three Seekers after Religion' (1910), who searches for the true religion in the established institutions of learning and finishes by knowing of no church, the doctrine of which he cannot refute (GS II.3, 892–4). The scepticism of Germany's awakening youth, accordingly, is said to be without limit (GS II.1, 74). What Benjamin desires in a new religion, then, is not the sanctification of human life in general but the sanctification of the life of that alleged intellectual elite to which he takes himself to belong. The experience that his 'leaders' have of a spiritual life is no longer to be limited to those rare moments of happiness which they find in the reception of truly pantheistic poetry and art (GS II.1, 21–2). In desiring the cultic exaltation of the everyday, they desire a comprehensive and so to speak solid re-enchantment of their world.

Although it is the function of religious cult that Benjamin desires to be fulfilled, he cannot prophesy a new cult on its own. He needs a religious doctrine if the new cult is to work. The unhappy participant in secular life faces not only the disappearance of objects to which such terms as 'sacred' can be applied but also the possibility that the vocabulary of holiness as such is meaningless. Without a new religious doctrine, he has no prospect of endowing it with meaning. Such a doctrine, however, does not exist: 'we' do not know anything of the new god (GS II.1, 34). It is in this quandary that Benjamin turns to the hope for divine intervention and revelation. Grace ('Gnade'), he suggests, will recreate ('neu schaffen') the sacred and the unsacred (GS II.1, 73), thereby restoring a meaning to the word 'sacred' and enabling youth to choose the holy life they desire. Yet the author does not humbly await the revelatory effects of Grace. By his very writing, he tries to provoke revelation into being.

He tries to create or reinforce in his audience (and in himself) that attitude which he believes to favour the advent of revelation. This attitude is one that allows the desire for the sacred to intensify to the utmost degree. Youth, we read, 'trusts that the sacred and the damned will be revealed at that moment when its collective will to choose will be strained to breaking-point' (GS II.1, 73). This construction of a temporal coincidence of highest desire and fulfilment draws causal significance from the preceding appeal, reminiscent of both Kierkegaard and contemporary decisionism, to a choice that creates its very objects.[95] Surely, if the choice of the sacred creates the sacred (to be) chosen, the utmost will to choose also brings about, or at least contributes to, the revelation of the sacred. By attributing to the new youth in general his confidence that the intensification of the desire will provide the desired, Benjamin can reinforce in the congenial parts of his audience any existing magical or, to use Freud's term, 'narcissistic' convictions regarding wish-fulfilment. Those of his readers who hold any such convictions at all – and perhaps everyone does to some extent – and who furthermore share his desire for a post-traditional sanctification will be even more inclined to have this desire intensify. Adopting this attitude, which Benjamin believes to favour revelation, is not an obvious thing to do even for those who share the desire. First, the intensification of one's desire prior to actual fulfilment means suffering. Second, actual fulfilment must appear unlikely if Benjamin is right in presupposing 'a scepticism without limit' in youth. And third, magical or narcissistic beliefs concerning the way in which to acquire what is desired may be widely held in Benjamin's youthful Wilhelmine audience; yet they will not prevail uncontested in anyone who is able to read – let alone, write – 'The Religious Position'.

Benjamin tries to counteract these reasons for a possible failure in two ways. He tries to make his audience and himself already benefit from the desired sanctification. And, to a certain extent, he attempts to rationalise the magical convictions on which he draws. Each of these two attempts to forestall failure forces him into an uncomfortable closeness to traditional religion.

The attempted rationalisation of the magical belief that increased desire and suffering will bring fulfilment is contained in Benjamin's appeal to Grace, which I have already mentioned. The appeal finds a parallel in the following, quasi-psalmodic, description of youth: 'Their need is the greatest and the help of God nearest to them' (GS II.1, 72).

Grace and the preparedness to help those in need can only be attributed to a benevolent person. By drawing on the idea of a personal and sympathetic God, Benjamin tries to counteract the notorious unreliability of magical wish-fulfilment. And indeed, the intensification of my desire and thereby my suffering is a suitable method to obtain what I desire if it can provoke into action a person who has both the power and the will to fulfil my desire. The almighty and, *cum grano salis*, supportive God of the Judaeo-Christian tradition is such a person. Benjamin's appeals to Grace and the notion of divine help repeat the partial rationalisation of magical beliefs contained in the 'old' religions.

This repetition renders his presentation inconsistent. Unrestricted scepticism certainly rules out the notion that Grace will (re-)create what we desire to exist. We know a lot about the 'new' God, moreover, if we know him to help those in greatest need. We then even know that the 'new' doctrine of the prophesied religion will contain the 'old' dogmas of God's personal essence and loving-kindness. Yet what matters in Benjamin's early writing and in 'The Religious Position' in particular is not intellectual consistency. What matters is that each idea and device, irrespective of their mutual intellectual compatibility, contributes to the overriding aim of (re-)creating the hope for religious excellence in both the author and his readers. Placed against the background of this aim, Benjamin's agreement with traditional theological doctrine remains a problem. The idea of a personal God, after all, is likely to conflict with some of the intellectual reasons that led the addressed 'elite' to reject traditional religion. The presupposition of God's personal essence, however, remains implicit. Since only an analysis of the manifesto allows one to recognise its hidden traditionalism, 'The Religious Position' can undoubtedly work in favour of an edification which those benefiting from it experience as untraditional or even revolutionary. Finally, 'new' religions are not created from nothing. Some use of traditional belief is therefore unavoidable.

It is the burdensome task of systematic theology to rationalise those religious myths on which cultic practice depends. Each of the major religions in the culture into which Benjamin was born conducts a precarious simultaneity with regard to the proximity and the remoteness of the divine. However comfortably they move among humans, the Olympic gods have their proper place and home high up on a mountain on which no Greek would set foot. Although strongly engaged in Israelite politics, Yahweh takes great pains to ensure that the Israelites will not climb Mount Sinai on which he descends. And the Christian God, though incarnate in Christ, is transcendent. What is more, any religion which aims at the exaltation of human life by providing it with a superhuman meaning – that is, perhaps, religion *tout court* – must have recourse to such simultaneity of proximity and remoteness. Benjamin, who in his mature work would paradoxically combine proximity and remoteness in the notion of aura,[96] knew about this from very early on. In 1912, he writes: 'Religion is based on dualism [...]. A divinity which is everywhere, which we bestow on every experience and every feeling, is a desecration and merely apt to gild our feelings' (GS II.1, 22). A divinity close at hand, is no divinity, it cannot *sanctify* human life. As Benjamin knows, however, true religion cannot consist exclusively in dualism either. Albeit an 'enemy' of mysticism, it itself is based on the devout ('innig') striving for the union with God (GS II.1, 22, 31–4). A divinity out of reach cannot sanctify *human* life. In other words, for the desire for sanctification to be fulfilled it takes both the remoteness and the proximity of the divine.

For post-traditional attempts to establish a sense of sanctification, the remoteness of the divine is relatively unproblematic. Given the scepticism that Benjamin finds in himself and in his audience, the creation of a sense of God's closeness is by far the more demanding task. Benjamin tackles it in various ways. One of these consists in his presentation of the youth's confrontation with the Wilhelmine establishment as an ordeal ('Gottesurteil'). The doctrine of non-action, he emphasises, is close to the heart of the new youth. Yet the limitless scepticism of youth forces it to love the struggle. 'In the struggle, too, God can rise.' The struggles of youth 'are ordeals, struggles, in which this youth is equally prepared to accept victory or defeat because nothing matters except that the sacred [...] be revealed in them' (GS II.1, 74).

Sceptics have no chance of finding happiness in contemplation. It is not a notion of ordinary practice, however, which Benjamin opposes to the ideal of a *vita contemplativa*. Instead, he makes up for the loss of the cherished fruits of contemplation by presenting the action into which sceptical youth feels itself coerced in terms of a struggle from which God may rise. He thereby suggests that the confrontation of his readers with the establishment may provide that supreme happiness which, according to Aristotle's unsceptical belief, solely contemplation attains.[97] The plausibility of this promise depends on the truth of the assumption that God is somehow involved in that generational conflict which renders life uncomfortable in the bourgeois living rooms of Wilhelmine Germany. By characterising youth's struggle as an ordeal, Benjamin strongly suggests that this is the case. In particular, his use of 'ordeal' suggests that God may justify the youthful fighters in and through the very activity in which they are already engaged, namely confronting their fathers. As a result, Benjamin's readers can feel the desired sanctification of their lives to be imminent if only they experience their rebellion as justified. This, I take it, they certainly do. Benjamin's use of 'ordeal' can therefore make them feel close to their hidden God.

What Benjamin calls an ordeal, however, differs slightly but significantly from the historical practice which the word denotes. Ordeals were usually practised among the members of a community who shared the belief in a common god. The point of this magical practice, which, although it continued to be exercised in witch trials, was dismissed by Gregory IX in 1234, was to show whether or not a particular person had that god on his side. Ordeals were not, at least not primarily, used to convince their participants of the presence (let alone the existence) of their god.[98] To create a feeling of divine presence, pre-modern societies had less spectacular and less dangerous means at their disposal, the most important of which were cult and prayer. Benjamin, however, uses 'ordeal' to create a sense of the hidden God's closeness.[99] He uses the vocabulary of magic, that is, in order to achieve the aim of cult. After all, no established cult is at his disposal. In its deviation from actual magical practice as well as in its purpose of providing a feeling of contemplative depth, his use of 'ordeal' agrees with his recourse to 'magic' in 'On Language' (Section 1.2.2).

A second strategy to suggest the proximity of the divine consists in the treatment of the religion to come as already present. This device resembles the 'anticipation' of eschatological revelation, in 'On Language', by the positive affirmation of (L). In 'The Religious Position', the suggested presence of the future religion is of a symbolic kind and youth is its symbol: 'The youth which affirms itself *means* the religion yet to come' (GS II.1, 73). Religion is symbolically present in youth. Conversely, youth, according to Benjamin, is also 'in' religion. This is what he suggests in the pamphlet's very first sentence, of which I give a recklessly literal translation: 'The movement of the awakening youth gives the direction of that infinitely remote point in which we know religion.' The awkwardness of this sentence is significant. Benjamin does not speak of the direction *towards* ('Richtung *zu*') the point of religion, a preposition that would emphasise the distance between his 'we' and religion. Instead, he chooses the less usual construction of a genitive complement, thereby suggesting the closeness of religion, a closeness, which he then combines with the idea of infinite remoteness. Furthermore, since 'we' know religion 'in' an infinitely remote point – and do not know it, as would be natural to say, to be at (or 'an') such a point – 'we' seem already to be situated in that infinitely remote point of which 'we' give the direction. Benjamin's language imitates that simultaneity of proximity and remoteness on which religious cults rely. However, since he cannot draw on the belief in God's past revelation, he has to relate that simultaneity to religion, and cannot relate it, as traditional religious practice is able to do, to the divine. Yet whoever desires religion considers it the way leading to the divine. The presence of Benjamin's God is the paradoxically removed presence of a way leading to him; it is a second order presence, and absence, only.

In 'On Language', the 'anticipation' of revelatory fulfilment with (L), the use of the vocabulary of magic and the creation, around 'name', of an aura of meaningfulness solicit that feeling of mystery that the believer finds in God's presence. Just as the mystery of traditional religions, the mystery of self-communicative language is set off against the 'emptiness' of profane thought and life, called 'bürgerlich' ('bourgeois') by the author (144; *65*). Although Benjamin refers repeatedly to religious authors and texts, the agreement between his central thesis and the religious tradition is not marked as such. The alleged self-communication of language, by consequence, may look acceptable to any sceptic in search for religious edification who is not particularly interested in the pedestrian efforts of analysis. Most importantly, perhaps, Benjamin subjects his linguistic absolute to the very simultaneity of remoteness and proximity which he knows to be essential to religion and which he had already enacted in 'The Religious Position'.

No less than its more traditional predecessors, the linguistic absolute has to look close to humans if it is to provide them with a sense of sanctification. Accordingly, Benjamin represents self-communicative language as eminently worldly and human in what can appear to be the pantheistic, humanistic and modernist dimensions of his essay. If the linguistic absolute is to have a more than human impact, however, it must live up to its superhuman nature. Therefore the author counteracts, contradicts or complements with their opposite all of his apparently humanistic, pantheistic and modernist claims. Though explicitly stating the identity of spiritual and linguistic essence solely for human language, in the last analysis, he accords such identity only to the divine word. The (so-called) infinity of all human language, he eventually states, is of a restricted character, compared to the unrestricted infinity of the divine word (145, 149; *66*, *68*). The divine word, in turn, is more than the unity of 'translations' between other languages, for it differs in kind, through its creative powers, from all other languages (148, 157; *68, 74*). God's self-revelation or language may be founded exclusively on man and the language in man, but man, his spiritual essence and language are founded on God (147, 149; *67, 68*). Finally, if, as Benjamin argues, the instrumental character of actual human communication originates in the Fall of linguistic spirit ('Sündenfall des Sprachgeistes') (153; *71*), the presence of language in modern technology – Benjamin uses the example of a lamp (142; *63*) – is at best the presence of corrupted spirit. Pure linguistic spirit must reside beyond the brave new world of modernity if it is true that the use of means is essential to modern technology.[100]

Benjamin does not address the problem of how the remoteness and the proximity of the linguistic absolute relate to one another. This makes it inappropriate to consider 'On Language' a piece of theory or (even) theology. While systematic theology labours under the problem of God's (infinitely) remote proximity as it results from allegedly sacred texts, these texts themselves efficiently serve the cultic function of religious edification regardless of the intellectual inconsistency that they create in so doing. It is no more a problem for John (1: 1–10) than for Benjamin to juxtapose the identity of God and word and the resulting closeness of the divine on the one hand, and the difference between God and word on the other: The Word *was* God. And the Word was *with* God. And John tries no more than Benjamin to explain his juxtaposition of the incompatibles. To the purpose of edification, however, any attempt to provide such an explanation is superfluous or even detrimental. It is therefore appropriate to consider 'On Language' a piece of religious or quasi-religious prose. The essay is religious, and not only quasi-religious, to the same extent that the notion of an intellectual's private religion makes sense.

If we describe 'On Language' as a religious text, it becomes a possible object of sceptical rejection. The view, in contrast, that Benjamin secularises traditional religion is apt to lull the sceptical concerns of a congenial audience. As regards 'On Language', that view is based on a one-sided reading of the essay, which ignores those 'dualistic' features by which the author situates his linguistic absolute in due remoteness. It furthermore relies on a simplistic and false conception of religion in conveniently ignoring the latter's cultic function which, notwithstanding the remoteness or transcendent status which religious doctrine may ascribe to the divine, is directed at the real life of people in this world. And it ignores this function in spite of the fact that Benjamin does not, and indeed desires a new religion for the very sake of its fulfilment. Benjamin, that is to say, knows what those readers of 'On Language' who consider the essay an attempt at secularisation ignore, namely that any sanctification has to concern, and start from, the profane. Before revelation is believed to set in, there is nothing more profane for the originally agrarian communities of the 'old' religions than a thorn bush, or bread and wine. The same applies to lamps in industrial societies.

Each religion which emphasises the irreducible otherness of the divine, and Judaism and Christianity certainly do, encourages by the same token a 'negative theology', the rejection, that is, of all attempts to furnish the divine with some approachability. In 'On Language', Benjamin's suspension of (L) in the notion of the non-communicable agrees with the iconoclastic implications of negative theology. However, since religious institutions cannot renounce the belief in the approachability of the divine if they are to fulfil their function of furnishing human life with an appearance of sacredness, they also must fight what they encourage. The policy of the Roman Catholic Church to make it an 'article of Faith that Faith was not necessary for a true knowledge of God', to quote a contemporary of Benjamin,[101] is only one example of the desperate proportions which this fight can assume. The 'negative' adversary of the institution is therefore no less a participant in a religious confrontation. The assumption that the iconoclasm that Benjamin shares with negative theology is devoid of religious fervour is as absurdly inappropriate as the assumption of religion's unconcern for the everyday. There is no negative theology without positive religion. Only, the sceptic Benjamin has to (re-)invent the latter before he can realise its negative potential. This realisation in turn allows him to accommodate and domesticate his scepticism within a religious, or quasi-religious, realm.

Benjamin's letter to Buber of July 1916 (GB I, 326–7) confirms that the juxtaposition of incompatibles in 'On Language' can be understood as deriving from the aim of religious edification. The functional characterisation which Benjamin gives there of non-mediatory ('un-*mittel-bar*') writing corresponds to his thesis of linguistic self-communication. Considering such writing the form of expression appropriate to himself, he characterises it as aiming at 'the crystal clear elimination of the unsayable'. The iconoclastic evocation of non-communicability, however, also finds its counterpart in the letter. The affirmed manner of writing is to 'lead to what is denied to the word'. In 'On Language', Benjamin attributes to language as such the fulfilment of those aims which he pursues as a writer. These aims, moreover, coincide with the traditional purpose of religious edification, namely to lead to the divine ('ins Göttliche zu führen'). And it is the traditional mystical belief associated with that purpose which, if true, solves the contradiction in

Benjamin's twofold description of immediate writing. The unsayable will be eliminated once the union with God has established that knowledge which is denied to the human word.

Pascal's sceptic, who desires faith,[102] is in an awkward situation. How can he gain the belief in God that he believes to be worth gaining? Pascal advises him to perform the rituals of religion as if he already believed in God. Taking holy water, having masses said etc. will eventually bring faith. The author of 'On Language' is in a more awkward situation still. He despises as opportunistic and not truly religious any synagogue or church he could go to. Desiring to see the life of the non-established intelligentsia sanctified, he comes to resemble a person who follows Pascal's advice not in church but in the fields and forecourts of academia. In the lecture 'Science as a Vocation' of 1919, Weber harshly criticised the then religious or would-be religious youth for expecting the satisfaction of their spiritual needs from *Wissenschaft* or theory.[103] He insisted on what he considered the academic's duty to refrain from value judgments. There was no religion without the sacrifice of intellect, he claimed. The only place in modern society where this sacrifice could be made without the loss of honesty was the old church. The pretension at academic prophecy, in contrast, was an offence against both the demands of rationality and any true religious feeling. The corresponding phenomena in contemporary intellectual culture rested on deception or self-deception.

For anyone who shares Weber's ideal of *Wissenschaft* or who has some experience of what it means to live in a traditional religious community, it is easy to sympathise with these views. Yet even in Weber's presentation, they miss the point of, say, Benjamin's dilemma. Cohen's belief in a religion of reason notwithstanding,[104] a return to traditional religious practice may betray the ideal of rationality no less than the furnishing of intellectual discourse with set pieces drawn from traditional religious belief. And precisely if the agents of *Wissenschaft* ought to refrain from value judgments, it may be impossible to decide where the duty of academics, let alone intellectuals, lies. The exclusion of spiritual edification from modern academic discourse, in consequence, may well be a question not of duty or morality but of historical contingency or power. Those sharing Weber's taste probably do better not to condemn the inconsistencies of post-traditional edification on the ground that such edification violates the allegedly well-established norms of academic discourse.[105] By refraining from such condemnation, they may not only avoid an inconsistency on their own part, but also improve their chances of encouraging what Weber calls 'clarity'.

2
Literature

2.1 Religious decisionism and art

Like Kafka and other contemporary writers whose work is characterised by a similarly problematic concern with religion, Benjamin grew up in a family without any moulding religious commitment.[1] In Benjamin's case, however, we know that his secondary socialisation was apt to create a desire for religion. In the present section, I place his early writing against the background of that cult of *Bildung* ('education', 'culture', 'formation') which his teacher Gustav Wyneken developed before the First World War. Benjamin adopted Wyneken's *Weltanschauung* as an adolescent from 1905 onwards. In March 1915, after Wyneken's public support for the war, he broke with his mentor, but emphasised that this break expressed his lasting loyalty to the 'true' Wyneken.[2] Accordingly, Benjamin's writing after 1915 still responds in a number of ways to requirements which he first encountered in Wyneken's cult. This assessment, as I shall try to show, is true even of his doctoral thesis *The Concept of Art Criticism*.[3]

In a short autobiographical piece, probably dating from 1932, Benjamin presents the country boarding school of Haubinda which he attended between 1905 and 1907 as the place where the seeds for his later life were sown (GS IV.1, 435). In 1912, he characterises his encounter with Wyneken, who taught philosophy and German literature at Haubinda, as the decisive intellectual event of his youth, calling himself 'G. Wyneken's strict and fanatic pupil'.[4] Wyneken, the son of a Lutheran pastor, was an important and controversial figure in the contemporary movement for educational reform and outlived Benjamin by 24 years. Due in part to his advocacy of an exclusively historical teaching of Christianity, he faced considerable difficulties from the authorities.[5] Both his alienation from Christianity and his persisting interest in religion in general find expression in the studies that he undertook in the Manichean and Gnostic traditions and in the Jewish Cabbala.[6] In 1906, he founded the so-called 'Freie Schulgemeinde Wickersdorf', both a boarding school and the (spiritual) centre of the *Jugendkulturbewegung*, with which Benjamin identified and which, in Benjamin's view, was responsible for his reluctance to join the Zionist youth movement.[7]

The educationalist Wyneken advocated education for education's sake, using for this purpose a religious language and analogies between himself and Jesus Christ.[8] Hans Blüher reports that Wyneken attributed to Wickersdorf the function of the 'station of *homines religiosi*' who in the Middle Ages administered and taught spiritual values '*ad majorem Dei gloriam*'.[9] Accordingly, Wyneken sees the school not as a worldly institution but, with explicit reference to the Luther's conception of the church, as 'an invisible [...] "assembly of souls in one spirit"'.[10] His belief in the merely historical importance of Christianity notwithstanding, Wyneken holds that all true education is based on religion.[11] He bases his ideal of *Bildung* neither on any intellectually justifiable considerations nor, indeed, on the assumption of an anthropologically constant need for religion.[12] Authentic religion, for Benjamin's mentor, rests exclusively on the individual's heroic decision to live *as if* religion in general were true.[13] Religious decisionism, that is to say, was part and parcel of Benjamin's intellectual upbringing.

> But we act *as if* we knew that behind existence, invisible to us, redemption and beatitude were waiting for their hour to come. We crown ourselves with a holy will, we light the torches of a proud and courageous faith, and so we walk our way, without a goal yet certain of the direction. If a saviour of the world exists, we shall meet him; if he fails to come, we cannot but act in accordance with him. [Wyneken, *Schule und Jugendkultur*, p. 181]

For Wyneken, *Kunst* is the realm in which the attitude of religious decisionism is created.[14] Instruction in the visual arts, literature and music is therefore of paramount importance in his attempt to develop his pupils' will to religion. It is only the experience of art ('das Kunsterlebnis'), he states, which is able to vouch for the truth of religion independently of traditional faith.[15] *Kunst* thus becomes the cultic medium of religious edification for the religious decisionist, who doubts the revealed truth of Christianity. Wyneken's religious decisionism, we may say, goes hand in hand with the sacralisation of art.[16]

It was in the context of this art-centred religious decisionism that Wyneken shaped Benjamin's love of literature, awoke his interest in philosophy and developed a critique of contemporary culture which strongly impressed the pupil.[17] This critique is aestheticist, conservative and radical in character. It combines a reaffirmation of the traditional appreciation of *Bildung* with the unqualified rejection of self-interest and purposive action. Whereas Kant reserved disinterestedness for the exceptional case of the pure judgement of taste,[18] Wyneken, in line with the standard motif of materialistic degeneration in the conservative critique of culture, blames materialism and self-interest for the alleged corruption of contemporary society at large. 'Serve *Geist*' ('diene dem Geist'), Benjamin's mentor proclaims, ought to be the principle of *all* human action.[19] In order to inculcate such service in young people, he considers it appropriate to promote the cult of genius in Germany's reformatory boarding schools. A so-called 'canon of the saints' apparently was to serve the same end.[20] True *Bildung* or the unconditional service of *Geist*, Wyneken teaches, contribute to the redemption of the world.[21]

The human contributor to the envisaged redemption through *Bildung* is an allegedly autonomous youth still to be created under the educationalist's leadership ('Führertum').[22] With his reliance on an as yet non-existent agent, Wyneken responds to the discrepancy between the industrial realities of the Wilhelmine Empire and the values which he propagates. These values cannot be drawn from the reality of social life or from the existing *ethos* of any class: the *Bildungsbürgertum*, in particular, offers at best a parody of disinterested *Geist*. Therefore the justification of Wyneken's programme of cultural revival has to come from outside society. His religious self-presentation and his idea of youth culture[23] are two strategies with which he creates the appearance of such an outside. They attempt to dissociate his programme from both the requirements of empirico-political assessment and any suspicion that he may actually be pursuing his own interests as an educationalist.

To the same extent, however, that they are mere means to reduce the implausibility of a cult of *Geist* in Wilhelmine Germany, Wyneken's rhetorical strategies in principle leave open the option to conceive of *Geist* as closely related to the state. Wyneken combines for this purpose Hegel's idea of state with Schopenhauer's negation of individual life in favour of a celebrated metaphysical 'will'. The state, he argues, is the as yet most powerful manifestation of the spirit's instinct of self-preservation, that manifestation of the objective, super-individual will, that is, for which no sacrifice is too great. What matters most, he proclaims in 1913, is that there exists something like the state which manages to have people in their thousands overcome their self-interest by sacrificing their lives for that something's sake.[24] It is not surprising, then, that Wyneken supported the First World War, just as did most of the less eccentric advocates of *Bildung*.

Young Benjamin, however, felt betrayed. He had believed, and would continue to believe, in the necessity of an 'outside'. While breaking with his mentor in March 1915, he declares his determination to remain faithful to Wyneken's newly religious idea of youth.[25] This declaration, though important in view of the fact that the broad parameters of Wyneken's programme will continue to shape Benjamin's writing, can conceal the differences that exist between pupil and teacher even before March 1915. Benjamin both absorbs Wyneken's sacralisation of art and wants to rule out any, in Nietzsche's phrase, aesthetic justification of existence.[26] He both relies on Wyneken's programme and tries to cure its alienating effects, for the pupil's prophecy of a new religion not only agrees with the ideal of disinterestedness in the teacher's notion of *Bildung* but is also concerned with ending the moral suffering of those members of the Wilhelmine intelligentsia who like Benjamin himself embraced the literary and aesthetic variety of German *Bildung* wholeheartedly. The *literati*, he claims, are forced by society's rejection to adopt a life of pathological self-destruction; at the same time, he considers them that 'enslaved' group from which the new religion will spread (GS II.1, 28–9). Absurd as it may seem, Benjamin's talk of enslavement is significant. First, it may reflect a closeness between his understanding of '*literati*' and 'Jews'.[27] Second, it updates Wyneken's characterisation of youth as enslaved[28] by projecting it onto that social group into which Benjamin is in the process of growing. Third, it allows him to use the emergence of Christianity – or certain beliefs both Nietzschean and Christian about that emergence – as a precedent in order to lend some plausibility to his prophecy of a new religion.[29] And finally, the talk of enslavement shows how great Benjamin feels the aesthetes' need for a dignifying distinction to be. He considers society's denial of recognition to his co-aesthetes a humiliation that destroys happiness or even life. A post-traditional, religious distinction is well suited to make up for such a humiliation. It looks sufficiently great in size, and it is the only one that is obtainable, if obtainable it is, without or even against the establishment. This 'transcendental' character of the desired distinction is decisive since solely a distinction from the 'outside' will appear valuable to Benjamin's co-aesthetes. Their Nietzschean 'love of the most distant' ('Fernsten-Liebe') after all shows that they are not only rejected by society but also despise contemporary society on their part.

For considering their rejection by society degradation, Benjamin must believe the art-loving *literati* to be entitled to claim social recognition. He does so indeed. By means of a somewhat conventional reference to Schiller's notion of joy, he presents all art as essentially related to human happiness and implies that, although mankind in the form of Wilhelmine society has little regard for the bliss offered by the repudiated *literati*, their mission is in principle equal to the highest kinds of human activity or even superior to all the others (GS II.1, 17). He settles for the latter option when ranking those of his contemporaries who find fulfilment within society or social commitment ('im Sozialen') as not the best and lacking in depth (GS II.1, 26). The *literati*, by consequence, are not merely entitled to claim recognition but count as society's unacknowledged cultural aristocracy. The shrillness of this indirect self-adulation reflects the fear of one's own irrelevance, which is also detectable in Wyneken's elitist idolatry of *Geist* and in the related conservative concern with cultural decline in general.[30] Such fear must be the greater the younger the cultural critic in a patriarchal environment, or the more exposed he is to rejection on, for instance, ethnic or racist grounds. The remedy which Benjamin envisages is accordingly more radical than Wyneken's. The Benjaminian 'outside' differs from Wyneken's not only by the absence of any support for the Wilhelmine state but also by a questioning of the redemptive value which Wyneken accords to *Bildung*. For the Benjaminian 'outside', *Bildung*, high culture or art are not external enough. This shows Benjamin's desire for religion to be more radically religious, as it were, than Wyneken's. It also shows that long before his encounter with Marxism and his sharp attacks on the persisting German cult of *Geist*,[31] he sensed that *Bildung*, though presented as the solution by his mentor, might itself harbour a problem to be solved.

Most significantly, the dialogue on religiosity starts with the suggestion that the enjoyment of art excludes a good conscience (GS II.1, 16). Benjamin's aesthete no longer finds Schiller's untroubled joy in art; the suffering and 'consciousness of a proletariat' disturb his peace of mind and deprive his delight in art of honesty (GS II.1, 18). The neoclassical fusion of morality and art, which Wyneken somehow manages to incorporate into his heroic decisionism,[32] does not convince his pupil. Even before his enthusiastic reading of

Kierkegaard's *Either/Or*,[33] Benjamin states that art fails to provide a moral basis for life (GS II.1, 21–2). The unity of nature and reason (or spirit), with which Schiller (and the Romantics) tried to confront the Kantian chasm between duty and inclination can only be experienced in short-lived moments of happiness on which the moral life of a community cannot rest even if it were to be a community of aesthetes only (GS II.1, 31–2). Art may contribute to the overcoming of its separation from life by providing life-values ('Lebenswerte'), such as beauty, a sense of form, and feeling (GS II.1, 17), but only a new religion will achieve the desired spiritualisation of life and its all-too-human conventions (GS II.1, 29). The aesthete's 'culture of expression' is the highest of all cultures, yet without the basis of a new religion it floats in mid-air.[34] For Wyneken, it is true, *Bildung* and aesthetic education also rest ultimately on religion. His indulgence in a sacralisation of art, however, is relatively untroubled by the moral scruples which haunt his pupil.[35] It is therefore easier for Wyneken than for Benjamin to engage in a cult of art or 'Kunstreligion' while leaving the desired new, so to speak religious, religion in a state of unspecified vagueness. Wyneken, furthermore, can build on the religious experience of his childhood, no matter how much he detested his protestant upbringing.[36] For Benjamin, in contrast, each religious promise that he, in accordance with Wyneken's instruction, finds in art necessarily points to something of which he has no or very little experience. The 'fanatic' pupil's need to anticipate the desired new religion in a concrete manner is all the greater.

However, the actual impact of Benjamin's subliminal divergence from Wyneken's *Kunstreligion* is very limited. An important reason for this is that Benjamin's response to the moral problem that he finds in aestheticism continues to be shaped by Wyneken's basic assumptions. Any socially committed activity, Benjamin believes, confirms merely the modern degradation of work to a matter of technical know-how (GS II.1, 20). The aesthete, being a *Kulturmensch*, a man of depth and culture ('Kultur'), cannot therefore regain his lost honesty by a commitment to the progress of civilisation ('Zivilisation') and its allegedly unending chains of means-and-ends (GS II.1, 20, 25). Confronting social injustice politically necessarily fails to be an honourable objective because the pursuit of objectives is itself, as Wyneken teaches, the root of society's corruption. The loss of traditional religious belief, moreover, has deprived human suffering of all spiritual meaning: 'Man hat das Leid entgöttert' (GS II.1, 19). As a result, social commitment is devoid of metaphysical gravity ('Ernst'). The absence of metaphysical grandeur can disqualify social commitment for Benjamin only because he accepts his mentor's imperative 'serve *Geist*' as the principle of all action.[37]

In spite of his discontent with aestheticism, Benjamin also follows Wyneken in emphasising the importance of art as art. As I shall shortly show in more detail, Wyneken insists on what he presents as art's exclusive concern with itself in order to free art from those 'philistine' appropriations that exclude his own use of art as a substitute for religion. Similarly, Benjamin has his 'we' subscribe to a programme of 'l'art pour nous', which answers their desire for life-values (GS II.1, 17), *and* takes them to be justified in using the slogan 'l'art pour l'art' to exclude any non-aesthetes from a discussion of art's *raison d'être*: '*L'art pour l'art* is the last barrier which protects art from philistinism. Without it, every Tom Dick and Harry would discuss art's right to exist just as they discuss meat prices' (GS II.1, 16). Surely, if Tom Dick and Harry were actually expected to benefit from the Schillerian joy through art, which Benjamin evokes in support of his aesthetes' quasi-aristocratic status, it would make no sense to exclude them from discussing the value of art. The envisaged exclusion undermines Benjamin's own justification of the aesthete's superior status. His reference to Schiller's notion of joy is not merely a somewhat conventional one but conceals the fact that Benjamin actually rejects or disregards Schiller's attempt to provide a socio-cultural legitimation for the high value with which art is credited in Weimar Classicism.[38] Benjamin, that is to say, upholds the claim of the aesthete's superiority without honouring the reason that he, Benjamin, gives in its support. The claim of the aesthete's superiority therefore stands in need of justification even for Benjamin himself. Given both his neglect of the neoclassical legitimation of art and his general affirmation of a necessary 'outside', the required justification, for Benjamin, can only be of a religious kind.

In 'On Language', language fulfils that function which Benjamin in his earlier writing attributes to the prophesied new religion. If true, his notion of language provides the aesthetic way of life and the aesthete's elitist self-image with a religious basis and moral justification. First, in so far as any communication of spiritual essence 'in expression' (GS II.1, 141; SW I, 62) ultimately rests on the emanation of God's creative word, the aesthete's 'culture of expression' must indeed count as 'the highest'. Second, Benjamin's integration of the aesthete's concern with *Gestaltung* into an all-encompassing life of divine expression means that any discrepancy between art and (the true) life dissolves. And third, the aesthete's exclusion of himself from the sphere of civilisation and the writer's related construction of a linguistic absolute are themselves positive moral actions if (the assumption of) the communication, in language, of extra-linguistic contents is indeed, as Benjamin claims, the Fall of linguistic or human spirit. The sphere of civilisation, in the sense of the 1912 dialogue on religiosity, is then one of corruption. Given that the de-spiritualised suffering of the proletariat is part of that sphere, the erosion of the aesthete's moral integrity originates in the sinfulness of civilisation. At the same time, the aesthete appears as justified before God if only he upholds *Kultur* in the face of *Zivilisation*. And so does the thinker of the linguistic absolute, by way of his thought, and the writer of 'immediate' prose, by way of his writing. Disinterestedness, which Kant attributes to the pure judgement of taste and which according to Wyneken characterises true *Bildung* in general, supports in Benjamin's early writing the aesthete's interested belief in his own righteousness. Although Benjamin emphasises that all art falls short of perfect revelation,[39] he must accord eminent importance to art, for it is the *aesthete's* sanctification which is at stake.

In *The Concept of Art Criticism*, Benjamin explores the potential meaning of art in a post-traditional religion. More particularly, he responds with his thesis to the aesthetic desiderata created by Wyneken's religious decisionism. In his *curriculum vitae* of 1912, Benjamin reports that Wyneken's teaching at Haubinda lastingly shaped his concern with literature through the development, in the pupil, of critical aesthetic norms (GS VII.2, 531). Wyneken's treatment of literature and criticism in his *Schule und Jugendkultur* of 1913, however, shows that the objective validity of the 'norms' which Benjamin evokes was far from established. What Benjamin calls norms cannot have amounted to more than mere claims about art which permitted its use in a post-traditional cult.

Wyneken stresses that the comprehension of works of art presupposes that of their structure and intention ('Wille'), as well as a rigorous study of artistic forms and laws. He thereby defends his teaching against the objection that it destroys the immediate enjoyment ('unmittelbares Genießen') that can be found in the experience, untouched by theory, of the work of art.[40] He tries to strengthen this defence by a counter-attack on the contemporary reception of art, which he pictures as combining artistic ignorance with a maudlin sentimentality. His rejection of an 'immediate', non-theoretical, appreciation of art, however, is far from unqualified. While criticising an audience that does not undergo the apperception of the work of art as such, he finds in that latter apperception 'really an immediate happiness', namely 'the vision or adoration of beauty itself'. It is wrong to look in art for something other than art, but the adequate experience of art as art, Wyneken maintains, discloses a true higher reality ('eine wirkliche Überwirklichkeit') and thus coincides in function with what he calls the 'culture of ecstasy'. The adequate experience of art frees those who are to be educated in the knightly service of spirit from any dependence on faith by allowing them to *see* ('schauen'). Since what they allegedly 'see' is not only beauty itself but also the rule of the Holy Ghost, I am tempted to invert John 20: 29 here: blessed are they that have not believed, and yet have seen. The religious purpose to which Wyneken puts what he calls art itself means that it is neither immediacy, nor emotion or edification as such which are excluded by his 'artistic' attitude; it is only the immediacy, emotion and edification of those philistines who do not conform to the educationalist's heroic decision to live *as if* religion were true. What is needed, then, to free Wyneken's claims from the suspicion of utter subjectivity is a theory of art that shows that the intrinsic structure of the work of art is essentially related to basic religious concerns. Such a theory, Wyneken frankly admits, does not exist, at least not with regard to literature. As regards literature, the pioneering achievement of a truly creative spirit will first of all have to establish the capacity of 'authentic judgement' ('eigentliches Urteilen').

The author of *The Concept of Art Criticism* strives to accomplish this task. Benjamin construes Friedrich Schlegel's poetics as identifying in the structure of both the work and the idea of art an objective criterion of criticism. The notion of art as a *Reflexionsmedium* ('medium of reflection') – in which he combines, in accordance with 'On Language', self-reference with 'medium' – is essential to that construction. On the one hand, Wyneken's religious ambitions are honoured by the understanding of both reflection and *Reflexionsmedium* as the, ultimately, ontological absolute. On the other hand, Benjamin's initial attempt to construe 'reflection' on the basis of a merely formal and methodological notion of the absolute respects the distance between religious decisionism and traditional religious belief. Nominally at least, Benjamin discards Wyneken's celebration of intuition and ecstasy by distinguishing Romantic 'reflection' from the notion of intellectual intuition and by classifying it as the opposite of Plato's *mania*. In one of the most problematic claims of the book, however, he agrees with Wyneken's insistence on true immediacy: the knowledge given in reflection, he argues, is immediate.

In Part One, entitled 'Reflection', Benjamin aims at identifying the epistemological presuppositions contained in the Romantic concept of criticism, which he develops in Part Two under the title 'Art Criticism'. The overall title of the book presents the latter development as Benjamin's major concern. The adequacy of this presentation has been questioned. Menninghaus argues that the *whole* book focuses on the Romantic notion of reflection and suggests 'The Theory of Poetic Reflection in German Romanticism' as a more correct and appropriate title.[41] It is true that Benjamin considers the 'theory' of Part One essential not only to 'art criticism' but also to the other key concepts of Early Romantic poetics which he treats in Part Two, such as 'irony' or 'the work of art'. Menninghaus's claim, however, that these other concepts are definitely not of merely secondary importance is doubtful. In each of the three chapters of Part Two, Benjamin treats 'criticism' and arguably aims with his discussion of other concepts at the reconstruction of Schlegel's notion of criticism. This is a sufficient reason for referring to art criticism in the title of the book and confirms Benjamin's introductory statement that he examines the Romantics' general theory of art only to the extent that it is important for their concept of criticism (13; *117*). Menninghaus's critical comment on Benjamin's title is nevertheless helpful. It draws attention to the fact that Benjamin's focus on 'criticism' in Part Two may not be a necessary corollary of the discussion of 'reflection' in Part One. For Benjamin, at any rate, the focus on 'criticism' cannot result from the 'theory' of reflection for he holds that in Schlegel's mystical terminology each mystical term or key concept is oriented towards, or contains the germ of, the system (47; *140*). In Benjamin's view, that is, the 'theory' of reflection is encapsulated in each of Schlegel's mystical concepts. 'Criticism' counts accordingly as only one – though an important one, and the one actually discussed – among Schlegel's many mystical terms (44, 50; *138, 141–2*).

This means that we need a non-theoretical reason to explain Benjamin's predominant interest in 'criticism'. The desire for sanctification which took hold of Wyneken's pupil is such a reason, for it concerns not primarily the artist or writer but the life and work of art lovers, men of letters, or – critics. At the same time, Benjamin's emphasis on *Kunstkritik*, on the criticism, that is, also of art, allows him to accommodate his doubt that art, in any empirical sense of the term, will ever provide what he desires.

2.2 Literary self-reflection

2.2.1 'Medium of reflection'

In this section, I shall analyse Benjamin's introduction of *Reflexionsmedium* against the background of the general methodological problems contained in his thesis. I shall proceed in four steps. 1. I shall outline the great significance of *Reflexionsmedium* in the overall argument of the book. 2. I shall consider some of the philosophical and philological problems contained in Benjamin's use of *Reflexionsmedium* and relate these problems to his methodological statements in the Hölderlin essay. 3. I shall show that Benjamin's introduction of *Reflexionsmedium* oscillates between the promise of discursive argument and an affirmation of mysticism. 4. The section will conclude with an explanation of the motives for my subsequent discussion and an outline of its development in the remaining sections of Chapter 2.

1. As is evident from both parts of the thesis, *Reflexionsmedium* is Benjamin's means of relating 'criticism' to the notion of the absolute. In Part One, the treatment of reflection leads to the following claims:

- Reflection is a medium.
- Reflection moves in a medium.
- *Reflexionsmedium* is the most appropriate designation for the (Romantic) absolute.
- The Romantic absolute is not the Fichtean I but art.[42]

Still in Part One, Benjamin states that in Part Two the absolute of art is examined on the basis of these principles and that art criticism is treated there as reflection in the medium of art (40; *134*). In Part Two itself, Benjamin credits Friedrich Schlegel with the achievement of having secured the autonomy of art objectively from the side of the work of art, and not subjectively, as did Kant, from the side of judgment.[43] This achievement, according to Benjamin, provides criticism with an objective basis and rests on what he presents as Schlegel's theory of art as a medium of reflection, and of the work of art as a centre of that reflection. Criticism establishes ('darstellt') the relation of the individual work to the idea of art, which idea is defined as the *Reflexionsmedium* of artistic forms.[44] If, however, the appeal to 'the idea of art' is not to exclude the existence of objective, critical criteria in the work of art itself, the idea of art must be essentially related to the particular work. The reflective medium that is the idea of art, to put it differently, must be essentially related to that medium which is the essential reflection in the work. Accordingly, Benjamin considers the form of the work, first, (the expression of) the work's self-reflection and, second, a centre in that medium of reflection as which he defines the idea of art (71–3; *155–6*). A reliable concept of an essentially continuous, reflective medium is therefore crucial to the substantiation of that achievement which Benjamin attributes to Schlegel's poetics.

2. On the occasion of introducing the term *Reflexionsmedium*, Benjamin comments on the ambiguity of this expression in his note 60 (*63*). The term, he states, represents reflection as a medium, and it represents 'the medium in question' as that in which reflection moves. This ambiguity he takes to entail no lack of clarity since reflection, as the absolute, moves in itself. Reflection, here, is conceived on analogy to the notion of self-communicative language in 'On Language', for Benjamin's qualification 'the medium *in question*' (my emphasis) leaves no doubt that we are to think of reflection as that medium in which reflection moves. Language counted as a 'medium' of communication by virtue of communicating itself in itself; and reflection, which moves in itself, now counts as a 'medium' of reflection. In section 1.2.2, I argued that Benjamin's use of 'medium' is unclear or even paradoxical. In his note 60 (*63*), he tries to discard any worry about a lack of clarity by justifying the ambiguity of *Reflexionsmedium* as resulting from the absolute character of reflection. In his note 141 (*146*), however, he clearly concedes that the 'theory' of the *Reflexionsmedium* may be inalterably obscure and is, in actual fact, paradoxical. This concession notwithstanding, he attributes 'a peculiar fruitfulness' to the Romantic statements on art. And earlier on, he stated that his aim was to demonstrate the 'philosophical fruitfulness' of 'criticism' for the conception of art as a medium of reflection, to demonstrate, that is, the philosophical fruitfulness of what, according to his note 141 (*146*), is based on, and leads to, self-contradiction (44; *138*). Benjamin's support for the notion that art is a self-reflective medium contains an indirect affirmation of the paradoxical.

As far as I can see, *The Concept of Art Criticism* proposes no specific argument in support of the alleged fruitfulness of the paradoxical. (The reader is, of course, free to *believe* that the book as a whole, somehow, is such an argument.) It is well known, however, that the affirmation of paradox is one of the salient features in Schlegel's poetics during the *Athenäum* period.[45] It is a salient feature, that is, of what Benjamin considers the field of his study (14–17, *118–20*). The philosophical problem of Benjamin's project, the (re-)construction of an admittedly paradoxical 'theory', might therefore be its philological virtue. In actual fact, however, the interpretational value of his thesis is as questionable as its philosophical one.

Benjamin clearly indicates that the Schlegel of the *Athenäum* period neither uses the term 'medium' in connection with 'reflection' nor develops a theory of reflection – an explicit theory, that is (16–17, 37; *120, 132*). Indeed, there is according to Benjamin only one passage in Schlegel's Early Romantic writings which amounts to a sufficiently clear affirmation of the view that art is a *Reflexionsmedium*, namely the famous sentence of *Athenäum* fragment 116, which characterises romantic *Poesie* as having the potential to hover on the wings of poetic reflection.[46] And there is, again according to Benjamin, only one piece of critical writing which does justice to that ideal of criticism which agrees with the so-called Romantic theory of reflection, namely Schlegel's review of Goethe's novel *Wilhelm Meister's Apprenticeship* (14; *118*). Neither *Athenäum* fragment 116, however, nor Schlegel's Goethe review receive anything like a sustained interpretation or analysis in Benjamin's thesis. Instead, Benjamin relies heavily on Fichte and Novalis. And he relies on Friedrich Schlegel's posthumously published Cologne lectures (1804–5), called 'The Development of Philosophy in Twelve Books',[47] which differ from the latter's writings of the *Athenäum* period by their explicit treatment of reflection and the strong influence, in Benjamin's words, of the Catholic philosophy of restauration (16–17; *119–20*). Benjamin is aware of the fact that the discussions of self-reflection in Fichte and in Schlegel's Cologne lectures do not deal with the self-reflection of art but centre upon the concepts of subject, self-consciousness or I ('Ich'). They are in conflict, that is to say, with Benjamin's notion of artistic self-reflection as 'Ich-frei', 'ego-less' or unrelated to any conception of the subject, especially the Fichtean.[48] This means that Benjamin finds in his sources either statements concerning art and art criticism which are not part of an explicit treatment of self-reflection or he finds an explicit treatment of self-reflection which is in conflict with the notion that the self-reflection of art is ego-less. This philological misfortune affects the central theme of his book since he conceives art criticism as knowledge ('Gegenstandserkenntnis') in the medium of reflection that is art, as knowledge in a medium, that is to say, the reflection of which counts as 'Ich-frei' (62; *149*). The concept of art criticism of *The Concept of Art Criticism*, in consequence, is not contained 'in' Benjamin's sources in any unproblematic sense of the word 'in'. It is not part of their content, or something with which their statements are explicitly concerned.

Drawing on the wording of the Hölderlin essay, we may say that the concept of art criticism that Benjamin develops in his thesis is not 'aktuell vorhanden', or *actualiter* present, in German Romanticism.[49]

It may still be present there in another way. It may, as the traditional distinction runs, be contained not in the letter but in the spirit of Early Romantic writing. Benjamin both questions and ultimately affirms this distinction: 'one must not take it so lightly as Friedrich Schlegel and speak of the "spirit of system which is something completely different from a system," yet these words do lead to what is decisive.'[50] By presenting the ego-less reflection of art as the epistemological presupposition of Schlegel's notion of criticism, Benjamin suggests that he is providing, as regards 'criticism', a sound basis for such a traditional use of 'spirit'. He suggests that the notion of art's ego-less self-reflection is implied in Schlegel's statements on art criticism around 1800. Yet *The Concept of Art Criticism* does not contain a logical or semantic analysis of these statements and is in general, to put it mildly, relatively unaffected by logical and semantic concerns. Furthermore, as Benjamin acknowledges in his introduction, he excludes from the very beginning those psychological interests of the Romantics that run counter to the idea of ego-less reflection (13; *117–18*). Thus, however, it is Benjamin's abstraction from the subject-oriented concerns actually present in both Schlegel and German Romanticism in general which seems responsible for the prominence which the notion of art's ego-less reflection gains in the thesis. This notion, it would seem, is not presupposed in German Romanticism but guides Benjamin's selective treatment. If we, wrongly perhaps, think of Early Romantic writing as a system with a firm, functional coherence or an inter-dependence between all its parts and the whole,[51] we may say that Benjamin's selective treatment amounts to the loosening-up of that coherence, or to a 'Auflockerung der festen funktionellen Verbundenheit', such as he grants the commentator in the Hölderlin essay (GS II.1, 106; SW I, 19).

The philosophical and philological problems of Benjamin's approach seem to be related to his application of *Problemgeschichte* ('problem history'); Wilhelm Windelband, the founder of the south-western school of neo-Kantianism, whose *Geschichte der neueren Philosophie* Benjamin uses, was the major representative of this controversial method in the history of philosophy.[52] Dealing simultaneously with systematic and historical concerns, *Problemgeschichte* is of course in danger of not doing justice to either side. More importantly perhaps, the methodological problems of Benjamin's thesis are in agreement with the author's programmatic statements and methodological reflections in the period concerned. They are not accidental weaknesses but intentional features of his writing or, at least, features of which he approves. Why is it that Benjamin uses philosophical and philological approaches and at the same time risks failure on both accounts? The simultaneous reliance on, and distance from, philosophical and philological argument in the thesis is reminiscent of the introduction to his Genesis commentary in 'On Language'. As argued in section 1.3.2, this introduction implies the (denied) affirmation of the belief in biblical revelation, is an agreement with a self-referential reading of one of the Bible's most mystical passages, and includes both an affirmation of mysticism and an early instance of that quasi-biological language which Benjamin, in the thesis, uses in the context of 'reflection'. Should we, then, explain the philological and philosophical fragility of *The Concept of Art Criticism* in terms of a mystical approach, essential to the book?

3. Benjamin introduces 'Reflexionsmedium' as follows:

> Schlegels Begriff des Absolutums ist mit dem Vorstehenden Fichte gegenüber hinreichend bestimmt. An sich selbst würde man dieses Absolute am richtigsten als das Reflexionsmedium bezeichnen. Mit diesem Terminus ist das Ganze von Schlegels theoretischer Philosophie zusammenfassend zu bezeichnen und wird unter diesem Ausdruck nicht selten im folgenden noch zu zitieren sein. Es ist daher notwendig, ihn noch genauer zu erklären [...]. Die Reflexion konstituiert das Absolute, und sie konstituiert es als ein Medium. [36–7; *132*]

> Schlegel's notion of the absolute has now been sufficiently determined[53] in contradistinction to Fichte's. In itself, one would most correctly designate this absolute the medium of reflection. The whole of Schlegel's theoretical philosophy can be designated and summarised[54] by this *terminus*, and it will frequently have to be cited by this expression in what follows. It is therefore necessary to explain the expression even more precisely. Reflection constitutes the absolute, and it constitutes it as a medium.

With his insistence on correctness, precision and the necessity to explain the terms used, Benjamin clearly appeals to the virtues of discursive discussion. In the third sentence of the passage, however, he strongly suggests a mystical understanding of 'Reflexionsmedium'. As the 'summary' of an entire (theoretical) philosophy,[55] the term 'Reflexionsmedium' cannot be part of a discursively developed argument but conforms to what Benjamin discusses in the subsequent chapter of his thesis as Schlegel's mystical belief in the systematic potential of single terms. The use of the Latin *terminus* underlines the relevance of this discussion to the introduction of 'Reflexionsmedium', and so does the use of the verb 'zitieren' ('to cite', 'to quote').[56] Surely, a philosophy cannot be quoted in the literal sense of 'to quote'; what can be quoted in the literal sense are only the texts in which a philosophy is represented. Yet in conjunction with the conditional construction 'would designate', the impersonal construction of the second sentence indicates that Schlegel does not use the term 'Reflexionsmedium' in his texts, an indication which Benjamin makes explicit shortly afterwards. The statement that the whole of Schlegel's philosophy will have to be 'cited' by 'Reflexionsmedium' may therefore mean that that philosophy will have to be called forth by means of (or rather '*in*') 'Reflexionsmedium', or, as a speaker of German who shares the personifying implications of Schlegel's view of systems may say, that it will have to be 'herbeizitiert' ('summoned up') by this expression.[57] 'Reflexionsmedium', in consequence, may assume the character of a name. It may have to count as one of those names, by which, according to Benjamin, mystical terminology tries to call the system or, for that matter, the absolute (48–9; *140*).

4. Precisely because of these indications of mysticism, I propose to take Benjamin's appeal to the ideals of discursive argument more seriously than is frequently done. I do so, first, for heuristic reasons. Even if Benjamin's introduction of 'Reflexionsmedium' were unambiguous, it would of course not provide us with conclusive evidence as to the character of his thesis; an author may do things other than he suggests. If we suspect that *The Concept of Art Criticism* is a work of mystical literature, as indeed I do, we ought to engage in a detailed discussion of Benjamin's claims to discursivity in order to ascertain, and not only suspect, that the book is mystical. Second, I shall pursue such a discussion in the remaining sections of Chapter 2 because of the importance of the issue in question. Many readers of Benjamin are presumably inclined to accept Susan A. Handelman's ready concession that Benjamin's work 'does not have the logical rigor or consistency of systematic philosophy'; yet many readers are presumably also inclined to consider his work philosophical or even, in Handelman's phrase, 'deeply philosophical'.[58] It is a question worthy of discussion, however, whether or not such a use of 'philosophy' and its cognates conceals or

misrepresents the mystical or religious character of what it, more or less consistently, presents as belonging to the secular concerns of modern academia. The same holds for the question whether Benjamin, as Menninghaus suggests,[59] proposes in his thesis a theory of reflection in which Schlegel's mystical terminology figures only as one realisation of the medium of reflection among others or whether Benjamin's very own, central notion *Reflexionsmedium* is itself a mystical term. If the latter were the case, even Benjamin's doctoral thesis, which counts as an academic, or, at least, as his most academic piece of work[60] would be better characterised as an unconventional religious exercise. As a result, Raymond Geuss's hope that the theory of the Frankfurt School will be able to travel 'without the transcendental baggage' would be thoroughly misplaced with regard to Benjamin's work.[61] Before one may reach this conclusion and accordingly dismiss Benjamin's appeals to the conventions of discursive argument as misleading, it is appropriate to examine these appeals in their own right.

If the claim that 'Reflexionsmedium' is a correct designation of Schlegel's notion of the absolute is to be substantiated, it is obvious that Benjamin must provide an account of the latter notion which is both sufficiently detailed and *introduces* the notion independently of 'Reflexionsmedium'. In his introduction of 'Reflexionsmedium', Benjamin seems to suggest that this requirement has been fulfilled in the preceding parts of his thesis. In order to assess his appeals to the virtues of discursivity, we therefore need to examine whether his comparison between Fichte and Schlegel leads to a determination of the latter's notion of the absolute which suffices for a meaningful discussion of the application, to that notion, of 'Reflexionsmedium'. In the following analysis of that comparison, I shall first consider Benjamin's claims concerning the alleged immediacy of reflection in both Fichte and Schlegel and their allegedly differing conceptions of the infinity of reflection. (Sections 2.2.2–3) In section 2.2.4, I shall discuss Benjamin's claim that the two authors attribute different metaphysical qualities to the absolute; as we already saw, Benjamin argues that Schlegel, in contrast to Fichte, is concerned with a reflection not in self-consciousness but 'in the absolute of art'. This claim is made after Benjamin's introduction of 'Reflexionsmedium' but obviously qualifies for a specification of Schlegel's notion of the absolute. With the notion of an ego-less reflection in art, Benjamin prepares the ground for the discussion of 'art criticism' in Part Two. In section 2.2.5, I shall examine his conception of the work's essential reflection as a self-transcendence towards the idea of art, by which he presents the work of art as a prefiguration of criticism. Section 2.3.1 will be devoted to Benjamin's discussion of 'immanent criticism', and section 2.3.2 to my final characterisation of the critical writing which he advocates and practises in *The Concept of Art Criticism.*

2.2.2 Immediate knowledge

With the 'immediate thinking of reflection', Benjamin writes, 'the Romantics penetrate into the absolute' (33; *130*). While this wording can evoke the idea of conquering a holy shrine, Benjamin's discussion as a whole appears more sober. There is, he maintains, complete agreement on the question of immediate cognition between the Early Romantics and Fichte's epistemological stance of 1794; Fichte deviated from that common position only after 1794 by basing immediate cognition on 'intuition' ('Anschauung'), and no longer on 'reflection' (20–1; *122*). Surprisingly, Benjamin's discussion of the relation between immediacy and reflection in Fichte, which I shall analyse first, is by far more sustained than the corresponding treatment of Schlegel, to which I shall devote the second part of the present section.

1. Paragraphs 6 and 7 of Fichte's 'Über den Begriff der Wissenschaftlehre' (1794) provide Benjamin's textual starting-points. He characterises the relation between reflection and immediacy in 'Begriff' in two, markedly different, ways.

Claim (A) Fichte [dringt] auf das wechselseitige Durch-Einander-Gegebensein von reflexivem Denken und unmittelbarer Erkenntnis. [19; *121*]

Fichte insists upon[62] the constitutive interdependence of reflective thinking and immediate cognition.

Claim (B) [Fichte] bestimmt [...] die Reflexion als die einer Form und erweist auf diesem Wege die Unmittelbarkeit der in ihr gegebenen Erkenntnis. [20; *122*]

Fichte determines reflection as the reflection of a form and in this way demonstrates the immediacy of the knowledge given in it.

In contrast to claim (B), claim (A) is concerned neither with the determination of reflection nor with the idea that the knowledge given in reflection is immediate. With claim (A) Benjamin merely states his view that Fichte insists on the interdependence of immediate cognition and what Benjamin, vaguely perhaps, calls 'reflective thinking'. As we shall see, this view may indeed conform to important (and problematic) features of Fichte's philosophical project. For Benjamin's purposes, however, claim (A) does not suffice. Since he presents 'reflection' as central to the Romantic concept of criticism, which he takes in turn to be based on the view of the form of the work of art as (the expression of) reflection, he needs a determination of reflection – if not, indeed, a definition – which can, at least in principle, be related to aesthetic concerns. If claim (B) is true, such a determination can be found in Fichte's 'Begriff'. The assertion that Fichte demonstrates that the knowledge given in reflection is immediate is not the primary concern in claim (B); this assertion may only be Benjamin's interpretation of what he claims is Fichte's determination of reflection as the reflection of a form.

In the earlier passage of 'Begriff' on which Benjamin draws (§6, esp. pp. 66–8), the second in his presentation, Fichte is concerned with the difference between logic and that science of all sciences, or knowledge of all human knowledge, which he calls 'Wissenschaftslehre'. He tries to argue for the assumption, crucial to his whole undertaking, that logic is subordinated to *Wissenschaftslehre*. He proposes that the essence of logic consists in the abstraction from any content ('Gehalt') present in *Wissenschaftlehre*. This leaves him with the problem that the propositions of logic seem to consist of form only. If this were true, there would be no logic at all for any proposition must have both a form and a content. This is where 'reflection' steps in. Reflection, Fichte holds, turns what is form in *Wissenschaftslehre* into a content, which, as the content of *logic*, possesses the form of logical propositions. And he equates the form of logical propositions with a particular interpretation of what he calls, without explanation, 'the general form of *Wissenschaftslehre*'. He summarises his argument in the only sentence of §6 from which Benjamin quotes: 'That "action [...] through which the form becomes the form of the form as its content and returns into itself is called reflection"' (20–1; *122*).

This sentence is obscure because neither the sentence itself nor its context seems to allow an unambiguous explanation of the expression 'the form of the form as its content'. We can understand how the obscurity comes about. For establishing a mere difference between logic and *Wissenschaftslehre*, it would suffice to say that the logician abstracts from the content of human knowledge and, by reflecting on the form of that knowledge, makes this form the content of his thought.

Wissenschaftslehre, in contrast, would be concerned with both the form and the content of human knowledge. Yet if logic, as well as being different from *Wissenschaftslehre*, is to be subordinated to it, the logician's reflection cannot count as an enterprise independent of *Wissenschaftslehre*. Therefore, Fichte presents the abstraction in logic not only as one from the content of human knowledge, but also as one from the content of *Wissenschaftslehre*. And for the same reason he suggests that the logician's reflection is but a particular interpretation of what he calls the general form of *Wissenschaftslehre*. Logical reflection thus seems derivable from, or even encapsulated within, the 'form' of *Wissenschaftslehre*, with 'form' oscillating between the form of human knowledge, as one of the contents of *Wissenschaftslehre*, and a possibly elusive general form of *Wissenschaftslehre*. The obscurity of the sentence from which Benjamin quotes is the price that Fichte pays for upholding the superiority of *Wissenschaftslehre* over logic.

In §6, then, it is not Fichte's aim to determine, or even to discuss, the notion of reflection. He aims, rather, at upholding the status of *Wissenschaftslehre* as the supreme science of all human knowledge and offers, in the process of doing so, an obscure explanation of the word 'reflection'. The phrasing of this explanation can seem to support Benjamin's assumption that Fichte considers reflection as manifesting itself in the mere form of knowledge (72; *155–6*). Given the ambiguity that surrounds Fichte's corresponding use of 'form', however, it is not clear what this assumption amounts to. Without an explanation of the phrase 'the form of the form as its content', which Benjamin does not even attempt, Fichte's remark on 'reflection' cannot as of right count as a determination of reflection on which any proofs, for instance that of the immediacy of the knowledge given in reflection, could be based.

Fichte's purpose in the second passage of 'Begriff' on which Benjamin draws (§7, esp. pp. 70–2), the first in Benjamin's presentation, is to establish how *Wissenschaftslehre*, as a science, relates to its objects. The objects of *Wissenschaftslehre* are the actions of the human mind. Fichte considers these actions necessary because they conform to, and indeed are determined by, laws; they are, in this sense, 'form'. He also uses the singular, speaking of 'the object' of *Wissenschaftslehre* or 'the necessary action of intelligence'. His claim is that *Wissenschaftslehre* differs from its object by raising that object into consciousness. This process of a making aware is considered a reflection. More precisely, Fichte states: 'Through this free action, something, which is already in itself form, the necessary action of intelligence, is now received as content in a new form, the form of knowledge or consciousness, and therefore that action is an action of reflection.'

What is problematic in this statement is the phrase that characterises the necessary action of intelligence as being 'already in itself form'. Benjamin seems to interpret Fichte's 'the necessary action of intelligence' as referring to, or implying the idea of, the pure form of the mind (20; *122*). Fichte, however, explicitly maintains that the content and form of human knowledge are inseparably linked before reflection occurs. And the assumption that even the highest action of intelligence, the self-positing of the I, is expressed in a principle which possesses both a form and a content is essential to the whole project of *Wissenschaftslehre*.[63] We can therefore explain the problematic phrase by the fact that Fichte, when using the phrase, disregards the abstraction from content that, as he clearly states, must accompany what he calls reflection. What Fichte then means is not that the action of human intelligence is pure form but that it is reflected in *Wissenschaftslehre* only in so far as it is form in the broad sense of being 'necessary' or conforming to a law. In contrast to Benjamin's, this interpretation has the advantage that it may be compatible with that belief which is central to Fichte's writings around 1794, 'Begriff' included, namely the audacious belief in the possibility of a science of human knowledge which differs from logic by dealing not only with the form of this knowledge but also with its necessary or 'pure' content (whatever this may be).[64] In §7, the interpretation outlined finds further support in the fact that it leads to a question that Fichte actually discusses shortly after using the problematic phrase 'already in itself form'. How can *Wissenschaftslehre*, he asks, abstract from what is not necessary in the action of human intelligence if it is only the reflection of *Wissenschaftslehre* which makes this action conscious? How can one separate the necessary from the accidental in something one is not yet aware of? Fichte's answer is that it is the philosopher's genius ('Genie'), his 'sense of truth' ('*Wahrheits*-Sinn') or his indistinct feeling of what is correct ('dunkle Gefühle des Richtigen') which guides him in that abstraction which must accompany the reflection of *Wissenschaftslehre*. The 'science' *Wissenschaftslehre*, we may say, is guided by intuition.

Although this line of Fichte's thought is apt to support Benjamin's claim (A), Benjamin does not take it up. There are at least two reasons for this. If, as Benjamin argues, the Romantic conception of the work of art overcomes the subjectivist deficiencies in Storm and Stress poetics (71; *154*), the theory of reflection on which that conception is allegedly based cannot be presented in terms of an exercise of genius. Second,

Fichte's appeal to the philosopher's genius is insufficient to support claim (B), which Benjamin does not clearly distinguish from claim (A). What Fichte's appeal shows is not that the knowledge given in reflection is immediate but that Fichte, within the framework of his general assumptions concerning *Wissenschaftslehre*, links the reflection of *Wissenschaftslehre* to intuition. I shall consider in due course the importance of some of these general assumptions to Benjamin's concern with immediacy. Since he does not discuss them, however, I first continue with his explicit argument.

Compared to Fichte's intentions, Benjamin's interpretation of 'the necessary action of human intelligence' can be described as a formalistic reduction. On the basis of this reduction, Benjamin offers the following explanation of 'reflection': 'Reflection, then, is conceived as the transforming – and nothing but transforming – activity of reflecting upon a form' (20; *122*). The assumption that reflection transforms and changes the form upon which it reflects is in conflict with the context of Fichte's statement according to which, in reflection, the form is received in a new form, the form of knowledge or consciousness. Fichte is concerned with the moving of the form from an unconscious state to a state of consciousness. He expresses this change of states by the phrase 'is received in the form of consciousness', with 'the form of', here, being a pleonastic addition to 'consciousness'. If he wants to argue, as indeed he does, that the reflection of *Wissenschaftslehre* makes a previously unconscious form conscious, he cannot imply that the *form* changes in the process, that is, that reflection changes not only the state or condition of the form but also the form itself. One does not raise to consciousness what one disfigures. Benjamin's use of 'transforming', in contrast, must imply that reflection changes the form upon which it reflects and not only its state. Otherwise his discussion of immediacy and reflection in Fichte would either be irrelevant to the notion of criticism in his thesis, or he could not (have the Romantics) conceive the reflection that is criticism as the development *and* destruction of the form of the work (86; *164*).

Benjamin's explanation of 'reflection', quoted above, is noteworthy. He explains 'Reflexion' by 'Reflektieren', 'reflection' by 'the activity of reflecting'. Both the mode of that activity and its object are specified, but not the subject. The specification of both the mode and the object has recourse to 'form'. The object is a form, and the mode is exclusively a transforming. At a later stage of his discussion, he considers Fichtean reflection 'a connection between two forms of consciousness (the form and the form of the form or knowledge and the knowledge of knowledge) which pass one into the other and return into themselves' (21; *122*). Here, 'form' seems to realise itself by becoming 'the form of the form' (and vice versa). On its own, the passage does not perhaps require this interpretation. Yet the notion that there exists an *entelechy* in form is indispensable if there is to be such a thing as the germ of reflection in the work of art and its form (78; *159*). For the Fichte of 'Begriff', however, human knowledge can perfectly well exist without its reflection in *Wissenschaftslehre* ever occurring. This is precisely what his characterisation of reflection as a free action amounts to.[65] For Fichte, there is no *entelechy* built into human knowledge or its form which urges that knowledge to become *Wissenschaftslehre*, or 'the form' to become 'the form of the form'. There is, of course, a related *entelechy* built into Fichte's view of human existence. Men only realise their essence by the exercise of freedom, an exercise that comprises, among other things, reflection. Yet at this stage of Benjamin's presentation, Fichte looks as unconcerned with the issues of practical philosophy as is, allegedly, Schlegel.[66]

Neither in §6 nor in §7 of 'Begriff', then, does Fichte offer a clear determination of reflection as the reflection of a form that could serve as the basis for a demonstration that the knowledge given in reflection is immediate. The wording of Fichte on which Benjamin relies is either itself unclear or requires an interpretation different from Benjamin's if the context is taken into account. That his source lends little support to Benjamin's immediacy claim, however, is not the main problem. The main problem is that Fichte denies the assumption of an immediate reflection in the very passages on which Benjamin draws. In §6, Fichte explicitly emphasises the interdependence not of reflection and immediate cognition but of reflection and abstraction; and in §7, he explicitly states that reflection presupposes abstraction.[67] Abstraction is generally taken to be an intellectual process that excludes immediacy. Benjamin, too, in 'On Language' opposes abstraction to the immediacy of pure language and knowledge (GS II.1, 153–4; SW I, 71–2). It is therefore difficult to understand how he can lay claim to finding immediacy in what Fichte, in the passages under discussion, calls reflection.[68]

Benjamin is not interested in the meaning of the word 'reflection' in those passages of 'Begriff' that he uses: his interest is not a philological or historical one. Rather, his immediacy claims can to some extent be explained by his particular response to the radical epistemological formalism of contemporary neo-Kantianism. Neo-Kantian epistemology frequently aimed at a conception of what for Kant were the sensual data given to the intellect as originating within the latter's own form. Ernst Cassirer, whose Berlin philosophy class Benjamin most probably attended in 1915, comes close to equating the content of thought with the form of thinking when characterising the needs of contemporary epistemology as follows: 'We need not the objectivity of absolute things but rather the objective determinateness of the *method of experience*. [...] Therefore the real *content* of the object of thought, to which knowledge penetrates, corresponds indeed exactly to the active f o r m of thinking as such.'[69] Cohen's *Kants Theorie der Erfahrung*, to which much of Benjamin's early thinking responds, represents a major expression of post-Kantian formalism and Cohen's pupil, Paul Natorp, categorically rejects as obsolete the assumption that any content is prior and given to the intellect.[70] Speth has convincingly interpreted the aesthetic implications of Benjamin's Hölderlin essay against the background of neo-Kantian formalism.[71] As regards *The Concept of Art Criticism*, the influence of neo-Kantian formalism allows us to understand that Benjamin selects those two passages of 'Begriff', the wording of which can suggest that reflection is a process within form. That influence, moreover, may also be at work in Benjamin's silence over Fichte's insistence on both a content of knowledge that exists independently of, and is given to, the reflection of *Wissenschaftslehre*, and the abstraction from content as a necessary concomitant of reflection. Finally, and most importantly, the radical formalism of neo-Kantian epistemology provides the background against which Benjamin's attempt to substantiate claim (B) ought to be placed.

Benjamin tries to substantiate his claim (B) in the following passage:

> Fichte meint [...], eine unmittelbare und sichere Erkenntnis durch einen Zusammenhang zweier Bewußtseinsformen (der Form und der Form der Form oder des Wissens und des Wissens des Wissens), welche in einander übergehen und in sich selbst zurückkehren, begründen zu können. Das absolute Subjekt, auf welches allein die Handlung der Freiheit sich bezieht, ist Zentrum dieser Reflexion und daher unmittelbar zu erkennen. Nicht um die Erkenntnis eines Gegenstandes durch Anschauung, sondern um die Selbsterkenntnis einer Methode, eines Formalen – nichts anderes repräsentiert das absolute Subjekt – handelt es sich. Die Bewußtseinsformen in ihrem Übergang in einander sind der einzige Gegenstand der unmittelbaren Erkenntnis, und dieser Übergang ist die einzige Methode, welche jene Unmittelbarkeit zu begründen und begreiflich zu machen erlaubt. [21; *122–3*]

> Fichte holds that he can demonstrate an immediate and certain knowledge on the basis of a connection between two forms of consciousness (the form and the form of the form or knowledge and the knowledge of knowledge) which pass one into the other and return into themselves. The absolute subject, to which alone the action of freedom is related, is the centre of this reflection and can therefore be known immediately. The point is not the cognition of an object through intuition but the self-cognition of a method, of something formal – the absolute subject representing nothing else. The forms of consciousness in their reciprocal transition are the sole object of immediate knowledge, and this transition is the sole method by which that immediacy can be demonstrated and made comprehensible.

If we consider this passage heedless of neo-Kantian formalism, Benjamin would seem to be simply wrong. He would seem to claim that Fichte considers the absolute I as immediately knowable by virtue of reflection.[72] Yet Fichte in *Grundlage* starts his argument by distinguishing that particular reflection which is concerned with the absolute subject or the I's transcendental self-positing (the only reflection which Benjamin seems to admit) from what is called 'reflection' in 'Begriff' (the only text on the basis of which Benjamin introduces Fichte's notion of reflection). The reflection concerned with the I's transcendental self-positing, Fichte insists, does *not* make the I's self-positing a fact of consciousness.[73] The absolute I, in other words, does not enter the form of consciousness or knowledge as a content; it cannot be known as on object. Nor, however, can it be known immediately in its reflection upon itself; the I cannot come to consciousness in such a reflection since, as Fichte explicitly states, it is 'never immediately aware of its action'.[74]

The reading of Benjamin's immediacy claims against the background of neo-Kantian epistemology that I shall now present aims at two things. First, I intend to show that Benjamin's presentation of the absolute subject as the object of immediate knowledge is part of an auxiliary mode of expression which, in accordance with the notion of immediacy in 'On Language', aims at the evocation of that sphere of neutrality with regard to the notions of subject and object which he had postulated already in his programme of the 'coming' philosophy; Benjamin's evocation of the absolute, that is to say, implies a religious interpretation of those tendencies within neo-Kantianism which are directed at a 'scientific' transcription of post-Kantian Idealism. Second, I intend to show that Benjamin's discussion, in contrast to what we have seen so far, indeed finds some support in Fichte.

Benjamin considers (Fichte to hold) the absolute subject as immediately knowable since it is the centre not merely of reflection but of 'this reflection' (21; *122*). By 'this reflection' he refers neither to knowledge nor to the knowledge of knowledge but to the (reciprocal) transition between these two forms of consciousness. Reflection, that is to say, counts here as the transition between two elements of a series, to which Benjamin will add further elements later on in his book. This transitional sense of 'reflection' is not the only one in *The Concept of Art Criticism*, but it is the one on which Benjamin rests his case for immediacy in accordance with his view of language as a medium that, by virtue of its essential immediacy, transcends the subject-object dichotomy. A transition does not have an object. Nor does it have a subject in the same sense in which 'I' can be said to be the subject in 'I reflect'. Questions of the type 'Who reflects?' and 'What is being reflected?' do not apply to reflection in so far as it is transition. Why is it, then, that Benjamin considers the transition in question a reflection? And how can he continue to speak of a subject, albeit only an absolute one which is the 'centre', and apparently not the subject, of reflection?

As to the first question, it is helpful to refer to Cassirer's evaluation of series formation as essential to thought. The true meaning ('eigenste Bedeutung') of thinking, Cassirer argues in *Substance and Function*, lies not in the concern of traditional metaphysics with what a number of entities have in common but in the necessary transition of thinking from one element of a series to the next.[75] Benjamin may call the transition from knowledge to the knowledge of knowledge (and vice versa) 'reflection' because it coincides with what according to this neo-Kantian view makes the essence of thought. Two facts make this likely. Benjamin introduces his discussion of immediacy with the questionable assumption that the Romantics equate reflection and thought.[76] And he himself clearly tends at crucial points of his presentation to use 'reflection' for thought in general (27, 33; *126*, *130*). At least on one occasion, Benjamin even seems to blur the distinction between reflection and intuitive thought.[77] Additionally, Benjamin may consider the transition reflection because he holds that the reciprocal transition between the elements coincides with each element returning into itself. As argued above, the *entelechy* that Benjamin sees built into the elements in question is not part of Fichte's argument in 'Begriff'. Surprisingly, however, such an *entelechy* does figure in Cassirer's attempt at replacing (or complementing) traditional metaphysics with a theory of functional conceptualisation. Surrounding the word 'nature' with noteworthy quotations marks, Cassirer characterises the necessity in the transition of a series as that of a dependency between its two elements a and b which is unequivocally prescribed by the nature of the elements themselves: 'durch die "Natur" der Glieder selbst'.[78]

Benjamin's designation of the absolute subject as the centre of reflection is also reminiscent of neo-Kantian concerns. His comments, in a different context, on the following passage from Windelband shed some light on the expression 'the centre', which remains unexplained in his discussion of immediacy: 'Function without a functioning being is for him [Fichte] the metaphysical *Urprinzip* [...]. The thinking mind does not first "exist" and only then become aware of itself by virtue of any particular promptings, but it originates in the underivable, inexplicable act of self-consciousness.'[79] The anti-ontological thrust of Cassirer's theory of functional conceptualisation is already present in Windelband's insistence on the absence, in function, of any underlying being. Yet Windelband also marks the metaphysical potential of this very idea by calling it Fichte's metaphysical *Urprinzip*,[80] and by presenting the underivable act of self-consciousness as the origin of thought and *Geist*. In contradistinction to Fichte, Benjamin refers to such an origin in his comments not by 'I' but by 'the absolute', 'Indifferenzpunkt' ('point of indifference'), and, finally, 'the centre' ('Mittelpunkt') (39; *134*). 'Indifferenzpunkt' recalls that 'sphere of total neutrality' with regard to the object–subject opposition in which he located, in his programme of the 'coming' philosophy, the quintessence of pure knowledge or 'God' (GS II.1, 163; SW I, 104). The geometrical metaphor 'centre' accordingly conceives the origin of thought without resort to 'subject', 'self-consciousness' or 'I'.[81] Moreover, 'centre' severs the idea of origin from that of a temporal beginning in the past and thereby agrees with those accounts of creation, for instance in Meister Eckhart, Luther and Rosenzweig, which conceive of creation as a-temporal or persisting.[82] This agreement with a certain theology of creation is relevant in the present context, for Benjamin joins Windelband in according a metaphysical quality to the origin of thought. And he relates, towards the end of his discussion of Fichte, 'origin' to both the alleged immediacy of reflection and the world in general. It is the immediacy of reflection which he claims directs Fichte in his search for the origin and explanation of the world; what is more, the immediacy of reflection is even said to indicate that that origin may be found in itself, in that very immediacy, that is to say, which gives the indication (25; *125*). Not only is the knowledge given in reflection supposed to be immediate, but the immediacy of reflection is also both self-referential and revelatory. For Benjamin, the immediacy of reflection points to its own hidden nature as that sphere in which the *Ursprung* of the world may be found.

In his initial treatment of immediacy, Benjamin does not reveal the metaphysical and religious intentions that go along with his use of 'centre' but presents the absolute subject as the self-knowledge of a method (21; *122*). Fichte's 'absolute I' tends to shrink (or expand) into the very transition of which it is the alleged centre, for Benjamin claims that the absolute subject is immediately knowable by virtue of being the centre of reflection *and* that the transition which he considers reflection is the sole object of immediate knowledge. The absolute subject or 'centre', one must conclude, '*is*' the transition or reflection (and vice versa). The allegedly transcendental condition of consciousness '*is*' consciousness. Cassirer's wording in *Substance and Function* can be taken to come close to supporting the assumption of such identities. To Benjamin's 'absolute subject' corresponds what Cassirer calls the generating relation ('erzeugende Relation') of a series.[83] The quality or even the essence ('qualitative Eigenart') of that relation, Cassirer states, is to be maintained unaltered in the structure of the generated series. This can be interpreted as saying that the essence of the generating relation is somehow present in the structure of the corresponding series. However, Cassirer must certainly exclude any such interpretation if he wants to uphold the scientific and unmetaphysical outlook of his theory of functional conceptualisation. Benjamin, in contrast, can welcome the implication that the absolute subject '*is*' reflection – or, translated into Cassirer's terminology, that the generating relation '*is*' what it generates – for this implication corresponds to the equation of spiritual with linguistic essence (or language), which in 'On Language' he required theory (ambiguously) to uphold. Benjamin, we may say, turns against the anti-metaphysical intentions of neo-Kantianism in the very process of adopting its formalism.

Benjamin's discussion of immediacy in *The Concept of Art Criticism* agrees with the equation of spiritual with linguistic essence in 'On Language' also in status. In 'On Language', Benjamin with the postulate of (L) presupposed, but did not unambiguously state, the general identity of spiritual with linguistic essence. In the thesis, he implies, but does not assert, that the absolute subject is the transition or method of thought. We reach this conclusion because he considers the object of immediate knowledge, first, the absolute subject and, second, exclusively the transition between the forms of consciousness. Yet Benjamin does not explicitly draw this conclusion. The phrase 'the absolute subject representing nothing else' characteristically leaves the decision suspended whether the absolute subject only represents the self-cognition of a method or whether it should be considered – or *'is'* – the self-reflective method in question.

What Benjamin explicitly states, however, is that the forms of consciousness in their reciprocal transition are both the sole object of immediate knowledge and the sole method by which the immediacy of knowledge can be demonstrated and made comprehensible. By demonstrating and explaining the immediacy of knowledge, the transition that is reflection demonstrates and explains the very manner in which it, reflection, is known. Reflection demonstrates and explains its own knowability. If we replace 'demonstrates and explains' by 'knows' and furthermore assume that Benjamin is likely to treat 'knowability' as an equivalent of 'knowledge' – in 'On Language', after all, he treated 'communicability' as an equivalent of 'communication' (Section 1.2.2) – we may express his thought by saying that reflection knows itself. And given that reflection is presented as so doing exclusively by virtue of the very transition that it is, we can add that reflection knows itself *in itself*. Reflection, in the words of (L), 'is the "medium" [...] in the purest sense'. Yet immediate knowledge, too, is this medium, for in knowing reflection, Benjamin maintains, immediate knowledge knows the very manner in which it, immediate knowledge, can be known. Reflection and immediate knowledge merge in 'the medium'.

This train of thought conforms to Benjamin's claim (A), according to which Fichte insists on the constitutive interdependence of reflective thought and immediate knowledge. It does not conform to claim (B), for the assumption, in (B), that the knowledge given in reflection is immediate suggests that Benjamin is concerned with a particular knowledge that is known by virtue of a particular form of consciousness, reflection; he might thus be concerned, for instance, with logical knowledge as it results from the reflection upon the form of first order thought. What Benjamin is actually concerned with in his discussion of immediacy, however, is not an attempt to demonstrate that reflective knowledge in any such narrow sense is immediate. Rather, he starts from the assumption that the absolute subject is immediately knowable and proposes that we should think of this knowability in terms of that transition which he calls 'reflection' and, elsewhere, just 'thought'. To the extent that he takes reflection to consist in the transition between various forms of consciousness, he cannot assume any knowledge of a particular content that would be 'given' in reflection. All there might be, rather, is a knowledge concerning the presence (or being) of the generative principle. As he construes them, neither reflection nor immediate knowledge are concerned with any knowledge different from that of 'the absolute', with which they, moreover, tend to merge. They are as self-contained as is language according to (L). While in 'On Language' he concentrated on language as the manifestation, or rather the *being*, of the absolute, in *The Concept of Art Criticism* he locates this being in 'reflection' or thought in general and accordingly states that reflection, by virtue of being the absolute, moves ('bewegt sich') in itself (36, note 60; *189, note 63*). However, if there is identity of spiritual with linguistic essence, the two approaches amount to the same thing.

With this discussion of immediacy, then, Benjamin clearly prepares for his later suggestion that the (Schlegelian) absolute is best designated by 'Reflexionsmedium'. But why should we adopt his train of thought? He argues neither for his recourse to neo-Kantian motifs, which provide his discussion with an aura of academic respectability, nor does he justify his use of these motifs against what are at least the anti-metaphysical intentions of neo-Kantianism. Instead, he presents his train of thought as concerned with something that Fichte has demonstrated. If there is any argumentative support for Benjamin's idea of immediate reflection, we have to find it in Fichte.

Benjamin's concern with the immediacy of reflection is indeed more closely related to Fichte's philosophy than the philological criticism of Benjamin's interpretation of 'reflection' in 'Begriff' would suggest. There are at least three features of Fichte's *Wissenschaftslehre* project that are relevant.[84] First, Fichte holds that we somehow immediately know of the first principle and the I's transcendental self-positing, which that principle expresses. Second, the foundation of *Wissenschaftslehre* is to show that this somehow-immediate-knowledge is indeed knowledge and not an instance of wishful thinking or misguided intuition. The reflection at work in the foundation of *Wissenschaftslehre*, to put it more precisely, aims at demonstrating that Fichte's notion of the I's self-positing is as certainly true as is the existence of any knowledge. Third, in his attempt at establishing such a foundation, the philosopher is caught in a circle. Fichte openly acknowledges the circular structure of his project and accounts for it in different ways. In 'Begriff', he states that the philosopher presupposes those logical 'laws of reflection' which he then finds constitute *Wissenschaftslehre*, and that *Wissenschaftslehre*, accordingly, is 'not only the rule but [...] at the same time the calculation'. The latter statement comes particularly close to Benjamin's view of reflection as something in which a method merges with its own constitutive principle.

For the Fichte of 'Begriff', however, the circle in question means that the philosopher's reflection does *not* demonstrate the validity of *Wissenschaftslehre* or the existence of that system of human knowledge which shows the intuition of the absolute I's self-positing to be unquestionably true. All the philosopher can achieve is to establish the probability of its existence. For Fichte, that is to say, there is no knowledge concerning the absolute I which would be, as Benjamin puts it, both 'immediate and certain' (21; *122*). For Fichte there is at best the probability that such knowledge exists. He tries to come to terms with this, given his appreciation of absolute certainty, unpleasant result in two ways. First, he appeals to the philosopher's love of truth; if the philosopher only tries honestly and hard enough not to presuppose what he wants to find, he can happily content himself with the probability of an absolutely certain knowledge. Second, Fichte argues that the philosopher presupposes what he then finds not in his argument, but only in his presentation; the philosopher presupposes in his discussion of the first principle the rules of logic that he then claims to derive from that principle merely in order to make himself understood.

The circular structure of Fichte's struggles to establish the one system of human knowledge is a strong reason to reject the ambitious project of *Wissenschaftslehre*. Not that there is in principle anything wrong with a limited certainty, a merely probable truth or even with circularity as such. Yet if we get only that, we may just as well renounce all talk of an absolutely certain first principle, an *Urprinzip* and the like. We can do so more lightly to the same extent that we do not, as does Fichte, expect epistemology to provide man with a spiritual home of irrevocable security. Benjamin, obviously, is not prepared to take up this option. If anything, his insistence on an immediate and certain knowledge of 'the centre' displays the even greater urgency that Fichte's existential concerns have acquired for some thinkers of the early twentieth century. It also points to the precariousness of Benjamin's attempt to overcome what Georg Lukács termed modernity's transcendental 'Heimatlosigkeit' ('homelessness'),[85] for Benjamin tries to make a stronghold out of what is the vulnerable spot of Fichte's discursive struggles. He contracts, as it were, the circle in the foundation of

Wissenschaftslehre into one point, called 'reflection', in which the (implicit) identity of the generative principle with what it generates coincides with a knowledge that counts as both immediate and certain. This is surprising especially since each of Fichte's strategies to minimise the impact of the circle must be rather off-putting to Benjamin. Fichte's moral appeal to the philosopher's love of truth is incompatible with the ostensible anti-psychological thrust of Benjamin's discussion. And Fichte's defence which implies that the philosopher can simply presuppose on the linguistic level of his presentation what he does not at the same time presuppose on the logical level of his argument cannot appeal to someone who sees all linguistic theory under the obligation to promote the equation of spiritual with linguistic essence. Yet what happens in Benjamin's thesis with regard to Fichte is similar to what happens in 'On Language' with regard to Hegel. In the 1916 essay, the insistence on immediacy expresses the author's desire to reap the fruits of dialectics, according to which spirit is present and essence is appearance, without going through the process of dialectical mediation (Section 1.2.2). In *The Concept of Art Criticism*, Benjamin tries under the heading of 'immediate reflection' to draw spiritual benefit from the central methodological *problem* of Fichtean *Wissenschaftslehre* without, however, discussing that problem as such. We can respect as someone's existential decision the refusal to do without the hope for something absolute. Benjamin's suggestion, however, that that hope is fulfilled in the form of thinking as such stands unsupported.

2. The assumption of immediate knowledge is essential to what Benjamin presents as the Romantics' notion of reflection. (18–19, 25; *121*, *125–6*). Yet Benjamin seems to import the immediacy issue from his discussion of Fichte into that of the Romantics; at least he shows at no point that the Romantics actually consider reflection immediate. Instead he, first, opposes reflection as an absolutely systematic thinking ('Denken') to intuition, second, claims that it goes without saying that for Schlegel the immediacy of knowledge has to be saved in spite of the systematic character of reflection, third, offers as his only relevant piece of textual evidence a passage from Schlegel's Cologne lectures in which Schlegel, however, insists on the possible immediacy not of reflection but of thinking, and, finally, concludes that the Romantics penetrate 'with this immediate thinking of reflection' into the absolute.[86] In Benjamin's presentation, that is to say, 'thinking' steps in between 'immediacy' and 'reflection' to support the claim that Schlegel considers reflection immediate. The implication, however, that for

Schlegel 'immediate thinking' is or characterises reflection is wrong, and is so both obviously and significantly.

In the Cologne lectures, Schlegel rejects the notion that intellectual (self-)intuition is the source of knowledge on the ground that intuition necessarily fixes and reifies the living I (KA XII, 330). In so doing, he may misrepresent Fichte's concern with intuition, for in *Grundlage* intuition as such is neither fixed nor fixing but consists precisely in the undecided hovering of the imagination.[87] More importantly, Schlegel's rejection of 'intellectual intuition' in the Cologne lectures is only recent. The Jena lectures on transcendental philosophy (1800–1) work with a positive notion of intellectual intuition, and in the *Athenäum*, Schlegel had even declared intellectual intuition to be *the* methodological postulate of theory in general.[88] If this change in Schlegel's assessment of 'intellectual intuition' is of substantial importance, it weakens Benjamin's assumption that Schlegel's epistemological stance during the *Athenäum* period can be derived from the Cologne lectures (34; *131*). Even the Cologne lectures themselves, however, clearly contradict Benjamin's construction of that stance. It is not the case that in these lectures Schlegel conceives 'reflection' as the positive alternative to 'intellectual intuition'. Neither does he consider, as Benjamin suggests, reflection both immediate thought and knowledge. Well to the contrary, Schlegel denies outright that reason ('Vernunft'), which he introduces as the faculty of reflection (KA XII, 326), provides any knowledge at all. Reason counts as completely unsuitable for all theory and speculation (KA XII, 331 and 340). Schlegel even considers the faculty of reflection the centre of everything that prevents knowledge and disqualifies it precisely because of its reliance on the belief in the notion of thing ('Begriff des Dings'), 'the error of all errors'; reflection, that is to say, shares the fatal shortcoming of intellectual intuition, reflection, too, reifies the I and changes it into a thing (KA XII, 341). This critical view of reflection as reifying is far from new in 1804 or 1805; it even can be considered the dominant one in Romantic thought.[89] Benjamin, however, does not even mention it. We cannot explain this fact by a limited access, on Benjamin's part, to the relevant material,[90] for Schlegel's Cologne lectures, Benjamin's main source for what he presents as Romantic 'epistemology', clearly insist on the reifying nature of (the faculty of) reflection.

Schlegel's positive alternative to the reification of intellectual intuition is not 'reflection' but 'feeling' ('Gefühl') or 'spiritual intuition' ('geistige Anschauung'). Spiritual intuition is free of the reification implied in 'thing' (KA XII, 355). It perceives in any apparent object not Fichte's 'Non-I' but a spirit, a living – Buberian, as it were – Thou. This Thou coincides with the essence of the 'object's' existence and its 'inner meaning' (KA XII, 350). Thus *Gefühl*, in contrast to reason and reflection, achieves '*true knowledge*' (KA XII, 357). The true knowledge of *Gefühl* is immediate; it is the 'immediate perception of the inner meaning' or spirit in the apparent object (KA XII, 355–6). For Schlegel, then, it is not reflection but *Gefühl* which is immediate knowledge. It is immediate *knowledge*, however, only because it rests on the three cardinal virtues of Christianity, on Charity, Hope and Faith (KA XII, 356–7). In Schlegel's neo-Platonic account, the understanding of meaning depends on that communion or 'marriage' ('Vermählung') between the perceiving and the perceived spirit that he calls 'Liebe' ('charity' *and* 'love') (KA XII, 351). *Liebe*, however, is only a necessary and not the sufficient condition of true knowledge. On its own, *Liebe* merely leads to guesswork ('*Erraten*') with regard to the kindred spirits in the 'things' around us (KA XII, 356). If the striving for reliable knowledge is not to wane – and Schlegel characteristically equates the fulfilment of this condition with the achievement of knowledge – we need an infinite, absolutely inexhaustible hope for consent and unity among everything that exists. We find such hope in Faith. By being based on Charity, Hope and Faith, the spiritual intuition that is immediate knowledge is necessarily directed towards the infinite and divine (KA 358). It is the faith in a divine '*Ur-Ich*' which, if anything, for Schlegel solves the riddle of self-consciousness (KA XII, 337, 325). Schlegel may not, as Benjamin puts it, 'penetrate into the absolute', but he certainly pays respect to the divine by affirming an immediate knowledge to which religious belief is essential.

Schlegel's 'epistemological' use of the Christian virtues is closely linked to that affirmation of immediate knowledge which Benjamin misleadingly attributes to the Romantic notion of reflection. Without doubt, Schlegel's resort to Charity, Hope and Faith belongs to what Benjamin considers Schlegel's adaptation of ultraconservative Catholic philosophy. Benjamin suggests that the reliance on Christian philosophy is new in Schlegel and can 'easily' be separated from the latter's epistemological stance during the *Athenäum* period. However, since according to Benjamin this stance is exclusively transmitted in the Cologne lectures, the separation cannot be so easy after all (34; *131*). It would have to be achieved against Schlegel's openly religious construction of immediate knowledge in the only source available. At any rate, Benjamin's silence about the religious basis of Schlegel's affirmation of immediate knowledge means that we should reject *Benjamin's* separability claim as unfounded.

Schlegel's religious construction of immediate knowledge comprises a number of concerns that Benjamin made his own before and after the composition of his thesis. In the Cologne lectures, Schlegel offers the outline of a conception of 'language' that closely resembles the one attempted in 'On Language'. In addition to the issue of immediacy, the similarities include the assumption that every entity has a spiritual essence and that language is the self-revelation of spiritual essence (KA XII, 350). They furthermore include the use of the vocabulary of 'communication' ('Mitteilung') and the central importance of 'revelation' (KA XII, 355, 341). And finally, they include the applicability of 'language' to aesthetics. Schlegel calls the spiritual essence or inner meaning which, on the basis of Faith, is immediately perceived in *Gefühl*, 'beauty'. As a result, 'spiritual intuition' merges with 'aesthetic intuition' (*'ästhetische Anschauung'*) (KA XII, 355, 357). The aesthetico-religious intuition of the inner essence, however, is tied to the external *Anschauung* as the soul is tied to the body (KA XII, 355). Benjamin will work with a similar conception of beauty in 'Goethe's Elective Affinities'. And his combination of this conception with the Platonian idea of *anamnesis* both echoes Schlegel's recourse to that idea in the Cologne lectures (KA XII, 348–54) and anticipates the further development of 'On Language' in the Epistemo-Critical Prologue of the *Trauerspiel* study.[91]

Benjamin, then, is clearly interested in what Schlegel discusses in connection with his religious construction of immediate knowledge. Schlegel's application of 'immediate knowledge' to aesthetics is highly relevant to the overall conception of *The Concept of Art Criticism*. Benjamin's quotations from the Cologne lectures, though apt to conceal the impact of Schlegel's religious concerns, cover passages which figure within the latter's presentation of immediate knowledge as based on Faith. And Benjamin characterises the Cologne lectures as an example of ultraconservative Catholic philosophy. For all these reasons, we cannot take Benjamin to ignore the religious substance of that idea of immediacy which he links to 'reflection'. Though aware of its religious substance, however, he presents the alleged immediacy of reflection as an epistemological concept and not as an expression of religious belief. Uncommented upon and unsupported by any epistemological argument as it is, this presentation agrees with the non-theological introduction of the linguistic absolute in 'On Language'. It results in a lack of transparency and exposes Benjamin's thesis, with

McCole's phrase, to 'constant misunderstandings'.[92] Lacoue-Labarthe holds that Benjamin considered all means fair to deceive the academic institution which allegedly imposed the genre of a thesis upon him.[93] Yet the appeal to anti-academic subversivity does not go to the heart of the matter; the dissociation of central postulates from their own religious substance characterises not only Benjamin's published and 'academic' thesis but also the related private writing of, for instance, 'On Language'. The characteristic features of the discussion of immediacy in Benjamin's thesis derive not from a simple subversion of academic standards but rather from the attempt to present in an academic context an expression of religious hope which is *of its own accord* bound to minimise the impact of positive, religious belief. They are directed not against academia as such but against both the writer's and academia's separation (outside theology) from religion.

2.2.3 Intensive infinity

In 1797, Benjamin correctly maintains, Fichte considers unrestricted reflection unsuited for the explanation of self-consciousness because it produces an infinite regress.[94] The distinction between the I which is aware of itself and the I which is the object of this awareness is essential to the reflection of self-consciousness. Yet in order to draw this distinction, we need to be aware of the I which is aware of itself; we therefore must make the subject in the self-consciousness on which we reflect the object of a higher subject's awareness and so forth. In 1812, Fichte even holds that reflection, on account of inducing the subject-object distinction into self-consciousness, can seem to destroy reality; this appearance of destruction, he argues, disappears only if reflection is carried through to an end, if, that is, it is restricted or made finite.[95] Similarly, in 1806 he discusses the danger that the entire form of reflection may collapse into absolutely nothing ('in absolutes Nichts'), a danger, against which he urges the acknowledgment of the restricted, finite character of the reflected act of thought.[96] Thus it is either the finiteness of human thinking in general or the restriction of reflection that secures for Fichte the epistemic function of reflection.

The Romantics, according to Benjamin, replace the Fichtean necessity of restricting reflection with a 'cult of infinity'. For them, the infinity of reflection is allegedly not affected by the emptiness of an infinite regress but is saturated, full, or, in an expression which evokes the idea of happiness, 'erfüllt' ('fulfilled') (25–6; *125–6*). To Fichte's 'absolutely nothing' into which reflection threatens to collapse, they allegedly oppose the assumption that the infinity of reflection can be understood as the *Zusammenhang* ('nexus', 'interconnectedness') between everything that exists. Such an infinity, Benjamin argues, is not an 'endless and empty' course ('Verlauf') but an unrestricted intensity in which everything is, as he insists with a phrase from Hölderlin's Pindar commentaries, infinitely (precisely) connected: 'unendlich (genau) zusammenhängen[d]'.[97]

This contrast which Benjamin construes between Fichte and the Romantics resembles that between the two views of history in 'The Life of Students'. History taken as an 'empty' course of progress in the 'infinity' of time displays a lack of *Zusammenhang* and precision ('Zusammenhanglosigkeit', 'Mangel an Präzision') while messianic thought knows of a state in which history 'rests, concentrated in a focus' ('in einem Brennpunkt gesammelt ruht') (GS II.1, 75; SW I, 37). Since all history is concentrated in one focus, this state of perfection shares the inter-connectedness of everything with what in the thesis figures as Romantic 'reflection'. These parallels between the thesis and Benjamin's earlier embracement of messianism[98] may be accidental to the notion of reflection in *The Concept of Art Criticism*. Or they may indicate that the discussion of infinite reflection in the thesis is, and perhaps is even exclusively, an instance of religious self-expression.[99] The first option seems to agree with some of Benjamin's statements both in the thesis itself and in his correspondence.[100] The second option finds support in his allusions to an indirect or preliminary treatment of messianism in the thesis, and in his (problematic) attempt to separate Schlegel's notion of *progressive Universalpoesie* ('progressive universal *Poesie*') from the idea of progress.[101] These allusions notwithstanding, Benjamin's treatment of infinite reflection cannot exclusively count as an instance of religious self-expression if at least one of the following contentions is true: 1. Schlegel's concern with intensive infinity is not religious in nature. 2. Benjamin's assumption that Schlegel considers the infinity of reflection full is correct and helpful. 3. Benjamin presents an argument for the notion that the infinity of reflection is full which does not rest on religious assumptions. I shall now try to show that none of these possibilities actually holds.

1. Let us first consider Schlegel's idea of intensive infinity. At least from the still neoclassical sketch 'Von der Schönheit in der Dichtkunst' onwards, Schlegel praises as 'plenitude' the, in principle, infinite quantity of differences in the beautiful object and elsewhere (KA XVI, 27–8). The notions of harmony, self-sufficiency and inner completeness are to account for what makes the quantity in question a whole as opposed to a mere, 'empty' sequence of differences (KA XVI, 29–31).

'Harmony' is meant to cover the unity of the manifold on any level below the absolute maximum of divine universality; 'self-sufficiency' and 'inner completeness' stand for the manifestation of universality ('Erscheinung der Allheit') in the form of the beautiful object. Yet it is only the semblance ('Schein') of intensive infinity that pertains to the beautiful object, for, as Schlegel emphasises, real self-sufficiency can only be attributed to the Supreme Being. Beauty, in consequence, is but the image of divinity, and art only the latter's language (KA XVI, 30–1). Though this view of art is not an essential part of any traditional religion, it is clearly based on religious assumptions concerning the existence and revelation of a supreme being.

The idea of an intensive infinity remains in force in Schlegel's Early Romantic notion of *Poesie*, and so does the corresponding oscillation of 'art' or *Poesie* between the revelatory manifestation of divinity and the mere appearance of something divine. In *Athenäum* fragment 102, Schlegel equates *Poesie* with the infinite (KA II, 180). In the supplement to his Lessing essay, he argues that each work and each art intend the undivided unity of the One and All (KA II, 414). According to August Wilhelm Schlegel, the poet strolls in the light of revelation (KA II, 186), and in a 1799 note of his brother Friedrich, the representation ('*Darstellen*') of *Poesie* is said to turn in its highest power ('Potenz') into revelation (KA XVIII, 304). Yet if, as the latter maintains in his Goethe review, the transition to what is both supreme and most profound is by necessity a leap (KA II, 139), the highest power of *Poesie*, in which revelation occurs, may well be religion and no longer *Poesie*, at least not a *Poesie* of which religion is not an essential part. In fragment 11 of 'Ideen', Schlegel explicitly draws this conclusion: Only with religion does an eternally full, infinite *Poesie* exist; without it, there are for ever only novels or that idle playing about ('die Spielerei') which is called fine art (KA II, 257). This view from a later issue of *Athenäum* may help to explain Schlegel's conversion to Catholicism; at the same time, however, it only echoes his 1796 belief that divinity is the highest part of the beautiful (KA XVI, 31).

What is new in Schlegel's poetics of the *Athenäum* period is not the idea of intensive infinity but the historicisation of that idea. Schlegel no longer takes the manifestation of universality to be related to the beautiful object in a manner for which an ahistorical, neoclassical aesthetic can account. The intensive infinity in which Early Romantic *Poesie* adopts God's traditional attributes of oneness ('simplitudo') and plenitude is now located more and more within the generative formation ('Bildung') of art and culture. Yet this historicisation, in contrast to what a secular understanding of 'history' suggests, does not negate the religious substance of the historicised idea but returns, after the Enlightenment, to the older tradition of *Heilsgeschichte*. *Athenäum* fragment 222 accordingly dismisses as irrelevant to modern history and 'progressive formation' ('progressive Bildung') everything that is unrelated to the Kingdom of God (KA II, 201).

2. Benjamin's claim that Schlegel considers the infinity *of reflection* full is misleading. Schlegel neither considers intensive infinity a quality of reflection in Benjamin's sense, of reflection, that is, as distinct from intellectual intuition; nor does all reflection, for Schlegel, benefit from the intensive infinity of romantic *Poesie*. In *Athenäum* fragment 102, the 'inner intuition of the infinite' counts as the sense of *Poesie* (KA II, 180). The Goethe review deals with the 'beautiful soul's' self-reflection which, as the self-erosion ('Selbstaushöhlung') of inwardness, resembles that collapse into 'absolutely nothing' which Fichte fears in infinite reflection (KA II, 142). And in the Cologne lectures Schlegel explicitly names what for him makes the difference between empty and full reflection: only where the notion of the I and the concept of the world are one, is self-reflection not characterised by the spiritual emptiness of an infinite series of mirror images (KA XII, 351). Thus, however, it is neither reflection as such nor its unspecified 'infinity' which guarantees fulfilment for Schlegel but exclusively the identity between world and I. Infinite reflection is only full if the world '*is*' knowledge, if it is that centre or '*Ur-Ich*' in which 'all the radii of philosophy' converge (KA XII, 337). In his 'cult of infinity', Schlegel worships a transempirical subject-object or 'x' which contains, or even '*is*', the quintessence of all being and knowledge. He worships, that is, what Benjamin in 1917 called 'God' (GS II.1, 163; SW I, 104).

By discussing Schlegel's positive notion of infinity as one of reflection *tout court*, Benjamin gives an exclusively 'epistemological' picture of Schlegel's underlying metaphysical and religious assumptions. This procedure is especially unhelpful with regard to *Athenäum* fragment 116. According to Benjamin, this fragment represents Schlegel's sole both pre-Catholic and (relatively) clear affirmation of the notion that art is a medium of reflection (63; *150*). To support this claim, he refers exclusively to that sentence of the fragment which deals with the possible exponentialisation of poetic reflection, and which comes third in the following quotation:

1. It [romantic *Poesie*] can lose itself so much in the represented that one would think that characterising all sorts of poetic individuals

would mean the world to it; still, there exists as yet no form equally suited to the full expression of the author's mind; so that some artists who merely intended to write their novel finished by having represented themselves.

2. It alone, like the epic poem, can become a mirror of the entire surrounding world, an image of the age.
3. And yet, it, too, can most of all, and free of all real and ideal interest, hover on the wings of poetic reflection in the middle between the represented and the representing, and can raise this reflection again and again to a higher power and multiply it as in an endless series of mirrors. [KA II, 82–3]

In (3), Schlegel chooses for the characterisation of romantic *Poesie* the very idea of an infinite series of mirror images that in the Cologne lectures stands for the vain emptiness of reflection. This positive use of the idea does not rest on any notion of reflection established in the fragment, for Schlegel uses, but does not discuss, 'reflection', and he conceives of romantic *Poesie* as intensively infinite before using 'reflection' in (3). In (2), the mirror metaphor establishes the universal potential of romantic *Poesie* with regard to external representation. By the *ironical* evocation of chance ('finished by'; 'von ungefähr') in (1), Schlegel alludes to a connection between external and self-representation but does not attribute it to reflection. The irony may thus draw not on his notion of reflection but on the postulate of the oneness of world and I which the Cologne lectures, in contrast to *Athenäum* fragment 116, present in conjunction with a negative view of reflection. Indeed, each statement of the fragment that precedes (3) may already presuppose the fullness of romantic *Poesie*. In the absence of an explicit discussion of 'reflection' in the fragment, and in the presence of negative accounts of 'reflection' elsewhere in Schlegel, it is therefore reasonable to assume that the fullness that (3) evokes derives from the component 'poetic' in 'poetic reflection', and not from the component 'reflection'. For Schlegel, the infinity of romantic *Poesie* is full, but it is so *not* by virtue of the full infinity of reflection.

Benjamin's comparison between Fichte and Schlegel suggests that the latter's notion of reflection is Fichte's, except that Schlegel considers the infinity of reflection full. (3), however, is not concerned with reflection in Fichte's sense. Hovering between the subjective and the objective, between the representing and the represented, Schlegel's 'poetic reflection' transfers to poetics what Fichte in 1794 calls imagination ('Einbildungskraft') and considers the faculty of intellectual intuition, explicitly opposing it to reflection.[102] Schlegel's reflection is fittingly endowed with wings, with the traditional iconographic attribute, that is, of imagination. In contrast to Benjamin's, an interpretation of (3) as dealing with *Einbildungskraft* has the advantage that (3) can be considered compatible with fragment 102, according to which inner intuition, and not reflection, is the sense of *Poesie* (KA II, 180). Of course, any recourse to 'imagination' requires one to explain why Schlegel, in (3), uses the vocabulary of reflection. Two facts seem important to me in this regard. First, as borne out for instance by (2), Schlegel's ideal of romantic *Poesie* is less removed from a mimetic understanding of art than is sometimes assumed, and the imagery of 'mirror' and 'reflection' expresses such an understanding competently. Second, due to the philosophical connotations of 'reflection', Schlegel with his use of 'reflection' reaffirms the postulate contained in *Lyceum* fragment 115, according to which philosophy ought to be one with *Poesie* (KA II, 161), but he only reaffirms that postulate. The romantic *Poesie* of *Athenäum* fragment 116 is an ideal and Schlegel's presentation of it little concerned with the difference between descriptive and normative accounts of literature. Just as all literature, for Schlegel, 'is or ought to be [...] romantic' (KA II, 183), we may say that for him imagination 'is or ought to be' reflection. Interpreted in this way, Schlegel's use of 'reflection' in (3) complements the central claim in the 'Oldest System Programme of German Idealism', namely that the philosophy of spirit be an aesthetic philosophy.[103] The aesthetic *Darstellung* or representation of spirit, as we may interpret Schlegel's congenial postulate, is philosophy or 'reflection'.

3. Benjamin does not present a specific argument in support of the claim that the infinity of reflection is full. Neither does he directly argue for the view that the infinity of interconnectedness that pertains to Schlegel's notion of world is an infinity of reflection. Instead, he considers the notion that the infinity of reflection is full Schlegel's 'axiomatic presupposition', the value of which is demonstrable by its fruitfulness for the theory of art (31, 26; *129, 126*). Elsewhere, he uses not the scientific expression 'axiomatic' but the philosophico-religious expression 'metaphysical credo' to characterise the view of art as a *Reflexionsmedium*. Even with regard to this credo, however, he insists on the necessity of treating it in the context of his thesis as a scientific hypothesis, a treatment which for him includes the 'immanent' clarification of the credo (62; *149*). Yet the clarification that Benjamin may offer for Schlegel's axiom is certainly not immanent in any sense that would require an explicit and sustained analysis of the relevant

passages in Schlegel. On the contrary, Benjamin relies on Hölderlin, and he relies on the so-called 'schema of Romantic epistemology', which he construes in part without any regard at all for textual evidence, and in part by combining short and problematic references to Fichte and Novalis as well as to Schlegel (27–32; *127–30*). Of course, if we get clarification, we may be able to do without immanence.

This is how Benjamin appeals to Hölderlin:

> Hölderlin [...] schreibt an einer Stelle, an der er einen innigen, höchst triftigen Zusammenhang ausdrücken will: 'unendlich (genau) zusammenhängen'. Das Gleiche hatten Schlegel and Novalis im Sinn, indem sie die Unendlichkeit der Reflexion als eine erfüllte Unendlichkeit des Zusammenhangs verstanden: es sollte in ihr alles auf unendlich vielfache Weise, wie wir heute sagen würden systematisch, wie Hölderlin einfacher sagt 'genau' zusammenhängen. [26; *126*]
>
> In a passage where Hölderlin wants to express an intimate and most conclusive connection, he uses the phrase 'to be infinitely (precisely)[104] connected'. Schlegel and Novalis had the same thing in mind when conceiving the infinity of reflection as a full infinity of inter-connectedness: in it, everything was to be interconnected in an infinitely multifarious manner, systematically, as we would put it today, 'precisely', as Hölderlin puts it more simply.

Benjamin here refers to Hölderlin's commentary on an enigmatic Pindar fragment, the translation or adaptation of which is entitled 'Das Unendliche' ('The Infinite').[105] Hölderlin plausibly holds that Pindar is concerned with a prominent issue in Greek philosophy, namely the relationship between justice and prudence. Justice and prudence, Hölderlin comments, are related in a third by which they are infinitely and 'genau' connected. Given the title 'Das Unendliche', the otherwise unexplained 'third' probably refers to the infinite. Given, furthermore, that Hölderlin interprets (or replaces) Pindar's 'to emauton houtō periphraxas diabiō' by 'so mich selbst/ Umschreibend, hinaus/ Mich lebe' – which in turn may be rendered by 'thus circumscribing myself, living, I come out of my shell' – he seems to be talking about a self-experience which transcends the boundaries of the empirical person and the limited interests of her 'prudent' reasoning.[106] Yet Hölderlin does not tell us how to conceive the 'close' or 'precise' interconnectedness of justice and prudence in the infinite. Neither do his comments contain any indication that the evoked self-experience is reflection in any epistemologically relevant sense of the word. Thus, however, Benjamin's appeal to Hölderlin's 'genau' merely invests the idea of intensive infinity with that aura of exactitude that, it is true, at times also surrounds 'reflection'. Without a detailed interpretation, which Benjamin does not even attempt, Hölderlin's text is far too vague to clarify the sense of Schlegel's 'axiom'.

Benjamin's 'schema of Romantic epistemology' comprises different levels of reflection that unfold between the two poles *Urreflexion* and 'absolute reflection' (31; *129*). The 'schema' is important in the present context because according to Benjamin it is by passing through its different levels that the full infinity of reflection can be grasped: 'In a mediate manner, this interconnectedness [of full infinity] can be grasped from infinitely many levels of reflection by passing, step by step, in all directions through all the remaining reflections' (26; *126*). The understanding which Benjamin envisages here is reflective, for it is through a process of reflections that the full infinity of reflection can allegedly be grasped. As stated explicitly, the envisaged understanding is also a mediatory one. We have seen in section 1.2.2 that for Benjamin 'immediacy' vouches for the closeness of spirit. By now conceding mediation he may be honouring the religious requirement that absolute spirit must also appear remote; or he may be trying to accommodate the importance of discontinuity in Schlegel's concern with the comprehension and vivid representation of full infinity.[107] However, since he had introduced reflection as immediate knowledge and considered reflection a medium thanks to its continuous ('stetig') *Zusammenhang*,[108] the incorporation of 'mediation' may undermine the basic assumptions of his thesis. His discussion of the so-called schema of Romantic epistemology can be read as an attempt to ward off this danger. Yet, as I shall now try to show, it is better understood as a religious exercise in accordance with that sceptical messianism and methodological nihilism which Benjamin advocates in his so-called Theologico-Political Fragment. Just as 'On Language' concedes the possible incorrectness of (L), *The Concept of Art Criticism* concedes that its central 'theory of reflection' may be invalid.

Benjamin argues that mediation through reflections, which allegedly permits the reflective understanding of the interconnectedness of full infinity, does not in principle rule out the immediate character of that understanding. By his note 30 (*note 32*), he seems to suggest that his discussion of Fichte yielded the insight that each reflection is in itself immediate; the mediate comprehension of full infinity, he concludes, can therefore be considered 'a mediation through immediacies' (26–7; *126*). As it stands, this argument for the compatibility of 'immediacy' with 'mediation through reflections' is unconvincing. First, the alleged compatibility only exists if 'mediation through immediacies' amounts to what Hegel would call 'vermittelte Unmittelbarkeit' ('mediated immediacy'). Yet without a convincing account of dialectical mediation, which Benjamin does not offer in the thesis, and which in 'On Language' he ruled out precisely by his insistence on immediacy, (Section 1.2.2) 'mediated immediacy' is simply a contradiction in terms. Second, as argued in section 2.2.2, Benjamin rested his case for immediate reflection on the transitional sense of 'reflection' and has not, as he now seems to assume, demonstrated the immediate nature of reflections in so far as they are situated on only one cognitive level.

Understandably, perhaps, Benjamin does not rely on a substantiation of his compatibility claim. Instead, he has recourse to the 'virtual thinkability' – '[es] ist freilich virtuell [...] denkbar' – of an absolute, unmediated, immediacy in the reflective comprehension of interconnectedness. It is with such absolute immediacy, he maintains, that the *Zusammenhang* would comprehend itself in absolute reflection (27; *126*). 'Zusammenhang', here, potentially occupies both the subject and the object position of absolute reflection. Shortly afterwards, however, Benjamin treats absolute, non-mediated immediacy no longer as merely conceivable but as an actual quality of the absolute's self-reflection: 'As already indicated above, the absolute comprehends itself immediately in [...] reflection whereas the lower reflections can approach the highest reflection only in the mediation through immediacy' (31; *129*). The shift from virtuality to actuality remains unexplained, and the misleading clause 'as already indicated above' is even apt to conceal it. 'The absolute', moreover, now takes the place which *Zusammenhang* hypothetically occupied in the previous quotation. This suggests, as does the content of the previous quote, that Benjamin uses *Zusammenhang*, the central term in his presentation of intensive infinity, as a cipher for 'the absolute'.[109] This interpretation is also supported by his characterisation of the cohesion in *Zusammenhang* as 'systematic', given that he also states that for Schlegel the absolute was the system.[110] 'Zusammenhang' in *The Concept of Art Criticism*, then, is not a notion which explains the alleged intensive infinity of reflection independently of the postulate of absolute and absolutely immediate reflection. It is not part of any epistemological theory of reflection which may or may not lead to, but is itself not based upon, the assumption of absolute reflection.

Neither the absolute itself, moreover, nor absolute reflection can actually be approached by 'the lower reflections'. Without doubt, Benjamin repeatedly states or implies that this is possible, but he also affirms 'the dissolution of the proper form of reflection towards the absolute'. In the unrestricted expansion of reflection, he maintains, 'the thinking formed in reflection turns into formless thinking which is directed at the absolute'. Given that he had introduced reflection as an attribute of form, this means that the reflective attempt to approach the absolute results in the destruction of reflective thought (20; *122*). 'Of course this dissolution of the strict form of reflection is dissolution only as far as restricted thinking is concerned' (31; *129*). To the extent that Benjamin uses 'reflection' for thinking in general, he diagnoses here the destruction of all restricted or human thinking in the attempt to pass through the infinitely many levels of reflection in search for the absolute. The danger which we saw Fichte perceive in infinite reflection, namely that the entire form of reflection collapses into 'absolutely nothing', has actually materialised in what Benjamin presents as the Romantic notion of the full infinity of reflection. The opposition with regard to the nature of reflective infinity which he construes between Fichte and Schlegel therefore rests exclusively on a different interpretation of that dissolution which affects reflection in the human thinker's struggle to ascend the ever higher orders of thought. For Benjamin, Fichte's 'absolutely nothing' must somehow coincide with 'the absolute'; for ultimately he declares that the road to the absolute lies in the progressive dissolution, and not in the advance, of reflection.

Benjamin equates the dissolution of reflection with an increase in mediation and suggests that it exists only for finite thought; the absolute, in contrast, allegedly thrives in an absolutely immediate and intact self-reflection. Yet we are not given any reasons for assuming that there is such a thing as the absolute's integer reflection that could be approached by what for human thought is the decomposition of reflection. Benjamin's shift from the 'virtual thinkability' of absolute self-reflection to its presentation as an actuality strongly suggests that, for him, there is no theoretical reason for this assumption. According to the theologico-political fragment, it is only the Messiah who creates the relationship between history and the messianic.[111] In *The Concept of Art Criticism*, it is only the absolute's self-reflection that, if anything, clarifies the 'axiom' of the full infinity of reflection. Only if and when the decomposition of finite thought reaches that reflection will Fichte's 'absolutely nothing' turn out to harbour perfect fulfilment. Independently of this uncertain event, however, the gradual passing through the infinity of reflection is undistinguishable from the destruction of human thought. The hope behind the 'nihilism' that Benjamin advocates in the fragment is that somehow the complete destruction or 'total transience' of nature be related to the advent of messianic happiness (GS II.1, 204). The hope which inspires his representation of 'Romantic epistemology' in the thesis is that the destruction of human thought will somehow lead to a union, in thinking, with the absolute. Just as the sceptical 'nihilist' of the fragment cannot base his hope on history such as it exists outside its – 'virtually thinkable' – messianic transfiguration, the commentator on 'Romantic epistemology' finds no reason in 'restricted thinking' to believe that the envisaged destruction of human thought is anything other than just that.

From the perspective of secular philosophy, 'the dissolution of the proper form of reflection' amounts to the dissolution of Benjamin's attempted 'theory' of reflection; there can be no valid theory of reflection where reflection is non-existent or destroyed. A religious thinker, however, can experience the dissolution of his own 'theory' as his very mode of approaching the absolute. Benjamin seems to pay tribute to the requirements of secular thinking by apparently limiting the range of the dissolution which he concedes. Thus he explicitly only states that 'reflection' in the sense in which he introduced the term is inapplicable between third-level thought and absolute reflection. And he claims that there is, apart from absolute reflection, one other instance of truly immediate reflection, namely 'Urreflexion' (30–1; *129*). 'Urreflexion' both remains unexplained and oscillates between a reference to first order thought and one to what Benjamin considers the 'proper' and canonical form – or the 'Urform' – of reflection, namely 'the thinking of thinking' of second order thought.[112] His attempt at damage limitation, however, triumphantly fails. The reflective nature of what he chooses to consider first-level reflection, the thinking of anything, that is, and not solely the thinking of thinking,[113] is at least as questionable as that of third order thought. And the problem of ambiguity which he diagnoses from the third level onwards, and which leads him to concede a dissolution of the 'strict' form of reflection, actually affects already his conception of second order thought. If second-level thought is, as he argues, the 'self-knowledge' ('Selbsterkenntnis') of the mere thinking of something (27–8; *127*), first-order thinking is in both the subject and the object position of second-level thought. Even the *Urform* of reflection, then, carries an ambiguity of that kind which for Benjamin destroys the 'strict' form of reflection. His discussion of the 'Romantic schema', that is to say, is an utterly weak defence against the secular rejection of the attempted 'theory'. Any possible strength which it possesses can be drawn only from the mystic's hope that the very destruction of his 'theory' will result in a higher kind of knowledge.

In his discussion of the 'schema', Benjamin comes close to revealing its ontological implications. Each of the two cases of a both simple and truly immediate reflection that he admits, namely absolute reflection and *Urreflexion*, is said to contain 'the content of all reality, all thought'; and the two cases are said to differ from one another exclusively by the degree of clarity ('Deutlichkeit') involved. Absolute reflection is characterised by the maximum, *Urreflexion* by the minimum of that clarity with which their full infinity has unfolded (31; *129–30*). This construction, as already the use of the term 'gradweise' ('step by step') in the phrase 'by passing, step by step, [...] through all the remaining reflections', is reminiscent of the idea of gradation in scholastic ontology, on which Benjamin relied in 'On Language' (Section 1.2.4). The scholastic conception of all being as partaking in God would certainly allow one to explain how any piece of (self-)knowledge, as Benjamin implies in his understanding of *Urreflexion*, can be said to contain 'all reality'. Given the scholastic conception, the (self-)knowledge of any entity is a knowledge of something which partakes in God, which partakes, that is, in the quintessence of all reality. Given moreover the basic assumptions of medieval realism, such knowledge comprises the recourse to universal concepts that somehow pertain to the known reality itself. Most surprisingly, however, Benjamin dismisses the model of gradation as relatively unimportant by calling it, *in contrast to the theory of full reflection,* a subsidiary construction ('Hilfskonstruktion') for the 'logicisation' of a line of thought that the Romantics failed to develop with perfect clarity.[114] All of a sudden, the 'axiom' of the intensive infinity of reflection, which Benjamin set out to clarify, has grown into a full-blown theory that no longer seems to require any clarification, not even a clarification of its possible ontological and religious implications.

Benjamin knows – and can expect any attentive reader to notice – that this is not the case. He returns to the issue of full reflection in the last chapter of Part One. Under the title 'The Early Romantic Epistemology of Nature', and in accordance with Novalis's 'magical idealism',[115] he declares: 'Everything that is in the absolute, everything real, thinks [...]. All becoming-known of a thinking entity presupposes the self-knowledge of that entity' (55; *144–5*). Everything real thinks, just as every entity, in 'On Language', communicates. The Romantics can consider reflection substantive knowledge or full, Benjamin now clarifies, because they presuppose that already the first and most basic ('ursprünglich') material ('stofflich') thinking of first order thought is full (54; *144*). Thus, however, not only Benjamin's conception of second order thought is ambiguous, as we have already seen, but also his definition of first order thought as 'the mere thinking of something'; for 'something' refers to both the object and the subject in first-level thinking. What exists in reflection, accordingly, is not a subject's knowledge of an object different from it, but the reciprocal transition between, or the union of, subject and object: 'in the medium of reflection, the thing and the knowing entity merge into each other. [...] There is [...] no such thing as a subject's knowledge of an object' (58; *146*). The mere thinking of everything real, as conceived here, agrees with 'name' in pertaining to both human knowledge and the known, and knowing, object itself (GS VI, 12; SW I, 88). *Urreflexion*, therefore, is as problematic a notion as is 'name'.

Benjamin cannot account for '*Urreflexion*' within the anti-ontological framework of his initial discussion. He eventually concedes that the fullness of mere thought at the basis of *Urreflexion* is an axiom – the 'theory' of full reflection having vanished again – which is responsible for the turning of the methodological absolute into the ontological one (54; *144*). The absolute, then, counts no longer as a mere function without any underlying being. It is no longer merely that constitutive principle of a method, as which Benjamin evoked it in his discussion of Fichte, but itself a being, or rather, and at the same time, Being, the ground of all being. For the ontological absolute, I take it, is God; it is, in Benjamin's words, that in which everything real '*is*'.

The ontological turn in Benjamin's discussion reads like a performative affirmation of Heidegger's claim that notions of function are formalised notions of substance. At least, this turn supports the view that *Benjamin's* 'functional' notion of reflection is in actual fact a metaphysical and even a theological one. Similar to how, in 'On Language', the idea of self-communicative language was introduced without explicit resort to the theological assumptions of medieval metaphysics, however, the notion of self-reflection in the thesis is introduced in the anti-metaphysical vein of contemporary neo-Kantianism. Turning from Fichte to the Romantics, Benjamin insists on the 'axiom' of full reflection but refrains from openly addressing its ontological implications. It is only towards the end of Part One and under the title 'The Romantic Epistemology *of Nature*' (my emphasis) that he openly acknowledges the ontological character of the Romantics' notion of the absolute and thus, implicitly, the theological implications of the idea of infinite and full reflection. In a note of the same chapter, he concedes that the 'theory of the *Reflexionsmedium*' may be inalterably obscure and is, in actual fact, paradoxical, 'especially in the problem of *Urreflexion*'. Immediately, however, he seems to suggest that this concession has no, or only a very limited, impact on the Romantic theory of art. This will not be so if his later contention is correct that for the Romantics the same laws govern the knowledge of natural objects and the knowledge of the work of art that is criticism (65; *151*).

2.2.4 'Ego-less reflection' and art

In the present section, I shall examine the two remaining features, apart from 'immediacy' and 'intensive infinity', which according to Benjamin characterise the Romantic notion of reflection, the 'ego-less' nature of reflection and its placement in the 'absolute of art'. My first aim is to show that 'the problem of *Urreflexion*' actually affects what Benjamin tries to exempt from it, namely (his conception of) the Romantic theory of art. Since he construes 'ego-less reflection' as essential to the Romantic view of art, the 'paradoxical' character of the 'theory of *Reflexionsmedium*' does not affect what he presents as the Romantic theory of art only if it does not affect the notion of ego-less reflection (39–40; *134–5*). Benjamin neither defines nor explains this notion. In the following passage, however, he prepares the ground for its later use by severing 'reflection' from the Fichtean focus on the subject:

> Die Romantiker gehen vom bloßen Sich-Selbst-Denken als Phänomen aus; es eignet allem, denn alles ist Selbst. Für Fichte kommt nur dem Ich ein Selbst zu, d. h. eine Reflexion existiert einzig und allein korrelativ zu einer Setzung. Für Fichte ist das Bewußtsein 'Ich', für die Romantiker ist es 'Selbst', oder anders gesagt: bei Fichte bezieht sich die Reflexion auf das Ich, bei den Romantikern auf das bloße Denken, und gerade durch diese letzte Beziehung wird [...] der eigentümliche romantische Reflexionsbegriff konstituiert. [29; *128*]

> The Romantics start from mere self-related thinking as a phenomenon; it pertains to everything, for everything is self. For Fichte, a self belongs only to the I, a reflection, that is to say, exists solely and entirely in correlation with an act of positing. For Fichte, consciousness is 'I', for the Romantics, it is 'self', or, to put it differently: in

> Fichte, reflection is related to the I, in the Romantics, it is related to mere thinking, and it is precisely this latter relationship which is constitutive for the peculiar Romantic concept of reflection.

The Romantics dissociate the structure of self-reference in the absolute self-positing of Fichte's I – which Benjamin, here and elsewhere,[116] links very closely to 'reflection' – from the notion of the I. They find this structure in 'self' and attribute it to everything. Benjamin quotes a footnote from Fichte according to which '*Self* presupposes [...] the concept of the I; and all absoluteness contained in 'self' derives from that concept.'[117] Benjamin's quotation underlines the fact that in his argument 'ego-less reflection' requires a definition of the Romantics' notion of self that does not depend on 'I'. Yet we find no such definition in *The Concept of Art Criticism*, and the indications which Benjamin implicitly provides lead precisely into what he plausibly enough considers the obscurity of that 'theory' of reflection which he attributes to the Romantics.

'Self' is introduced in analogy to Fichte's 'I': For Fichte, consciousness is 'I', for the Romantics, it is 'self'. Since Benjamin is aware of the fact that Fichte considers the I's self-positing not consciousness but the condition of the possibility, if not the *Ursprung*, of consciousness, 'For Fichte, consciousness is "I"' should be interpreted as expressing Benjamin's tendency to equate the (no longer?) transcendental condition of consciousness with consciousness which also underlies his idea of immediate reflection (Section 2.2.2). Accordingly, 'Consciousness *is* the *Ursprung* of consciousness' would be a description of what Benjamin tries to find in Fichte's 'I'. In the context of Benjamin's overall argument, 'For Fichte, consciousness is "I"' results in an undecided hovering above the equation of the *Ursprung* of consciousness with consciousness, similar to that which in 'On Language' is programmatically required with regard to the identity of spiritual and linguistic essence (Section 1.3.1). Since he takes 'self' to relate to consciousness in the very same way in which Fichte's 'I' relates to consciousness, 'self', for Benjamin, tends to merge with both the origin of consciousness and consciousness itself. 'Self', if you like, is a name for that consciousness or absolute thinking which finds its origin in itself.

'Selbst', moreover, is a condensed substantivisation of 'Sich-Selbst' in the expression 'Sich-Selbst-Denken' ('self-related thinking'). Benjamin uses 'thinking of thinking' as an equivalent of this expression, and he also uses, in addition to 'self', the phrase 'mere thinking' to describe what according to him takes the place of Fichte's 'I' in Romantic thought (29, 40; *128, 134*). In Benjamin's use of 'self', that is to say, thinking and the thinking of thinking become indistinguishable. As a result, the idea of ego-less reflection is affected by the same lack of clarity that characterises the relation between first and second order reflection in Benjamin's 'schema of Romantic epistemology' (Section 2.2.3). It shares, as we may put it, 'the problem of *Urreflexion*'.

'Ego-less reflection' also resembles '*Urreflexion*' in that its presentation agrees with Benjamin's unconvincing attempt at steering clear of ontology. Taken as an appeal to Husserlian phenomenology, Benjamin's qualification of self-related thinking by 'as a phenomenon' emphasises the irrelevance of ontological concerns. The repeated use of 'mere' ('bloß') in conjunction with 'thinking' and 'thought' equally signals that the Romantics' 'self' is not meant to be an ontological concept. Benjamin underlines these hints by claiming that the Romantics drop the ontological determination of 'reflection' which he now – in contradiction to his initially formalist interpretation of Fichte – finds in Fichte's (alleged) placement of reflection in the absolute I's positing ('Ursetzung') or being ('Ursein') (29; *128*). For at least three reasons, however, this attempt to dissociate the Romantic 'self' from ontological concerns fails to convince. First, the assumption that Fichte's 'I' is an ontological concept contradicts not only Benjamin's initial interpretation but also Fichte's own beliefs until at least 1797.[118] If we nevertheless suspect that the Fichtean 'I' is essentially related to 'Ursein', we ought to do the same with regard to what Benjamin presents as the Romantic notion of self. Second, the support that Benjamin may derive from Husserl is double-edged at best. In 1913, a number of years before the composition of *The Concept of Art Criticism,* Husserl publicly renounced his earlier rejection of 'the pure I' and started to associate phenomenology explicitly with the subject-oriented assumptions of transcendental philosophy.[119] The view that this development should be completely unrelated to his notion of phenomenon before 1913 is implausible. Third, and most important, Benjamin's very own comment on the Romantics' alleged divergence from Fichte is an ontological statement: 'everything is self'. Accordingly, he contradicts his earlier claim of an *unqualified* absence of ontological concerns in Romantic 'reflection' by stating that 'Romantic thinking sublates being and positing in reflection' (29; *128*). The context of this statement leaves no doubt that 'aufheben' ('to sublate'), here as in Hegel, implies a positive element of retention.

I now turn to Benjamin's first use of the phrase 'ego-less reflection'.

> Erst mit der Reflexion entspringt das Denken, auf das reflektiert wird. Darum kann man sagen, jede einfache Reflexion entspringe absolut aus einem Indifferenzpunkt. Welche metaphysische Qualität man diesem Indifferenzpunkt der Reflexion zuschreiben möchte, steht frei. [...] Im frühromantischen Sinne ist der Mittelpunkt der Reflexion die Kunst, nicht das Ich. Die Grundbestimmungen jenes Systems, das Schlegel in den [Kölner] Vorlesungen als System des absoluten Ich vorlegt, haben in seinen früheren Gedankengängen ihren Gegenstand an der Kunst. Im also verändert gedachten Absolutum wirkt eine andere Reflexion. Die romantische Kunstanschauung beruht darauf, daß im Denken des Denkens kein Ich-Bewußtsein verstanden wird. Die Ich-freie Reflexion ist eine Reflexion im Absolutum der Kunst. [39–40; *134*]

> Only in reflection does the reflected thinking originate. It is therefore possible to say that each simple reflection arises absolutely from a point of indifference. One is free to choose the metaphysical quality which one wishes to ascribe to this indifference-point of reflection. In an Early Romantic understanding, the centre of reflection is art, not the I. The fundamental determinations of that system which Schegel propounds in the [Cologne] lectures as the system of the absolute I concern, in his earlier thought, art. In this differently conceived absolute, there is another reflection at work. It is basic to the Romantic view of art that the thinking of thinking is not conceived as self-consciousness of the I. Ego-less reflection is a reflection in the absolute of art.

Surprisingly, Benjamin appeals to logic to support the first proposition of this passage. Reflection, he claims, is logically prior to the thought upon which it reflects,[120] for since reflection is the form of thought, the thinking that is being reflected upon is logically impossible without reflection; hence 'Only in reflection does the reflected thinking originate'. The claimed logical priority, however, cannot support the genetical point that Benjamin rests on it. *L'Éducation sentimentale* presupposes the existence of French, but this does not make French the origin, or cause the existence, of Flaubert's novel. Reflected thinking does not necessarily spring from reflection even if it should presuppose reflection. Benjamin's argument is telling on a different score. It shows us that the conception of *any* thought as (being of the form of) self-reflection is part of that idea of absolute reflection that he links to art.

As a negative term, 'ego-less reflection' tells us only how not to think about reflection. In principle, the integration of 'reflection' into a theory of art might remedy this lack of positive information. Yet as Benjamin uses 'art' in the present context, the term does not contribute to a specification of 'reflection' but comes to share the obscurity surrounding 'reflection' in his earlier observations. In an Early Romantic understanding, he claims, art is the centre, alias the indifference-point, alias the absolute, of reflection. 'Art' thus occupies that position which before was held by 'absolute reflection' *and* by 'reflection'. Art, moreover, is that *in* which (ego-less) reflection 'moves'. As reflection *and* that in which reflection moves, art is presented here as a *Reflexionsmedium* in accordance with Benjamin's comments on this term in his note 60 (*note 63*). In the relation which he establishes between 'ego-less reflection' and 'art', then, we do not find an explanation of either concept, but the proposal to conceive (the Romantic concept of) art as the absolute, and, indiscriminately, as reflection.

In Early Romanticism, 'art' and *Poesie* certainly benefit from the aura of 'the absolute'. It is also true that we find, in conjunction with *Poesie*, both a proliferating use of expressions which contain 'self' and, though much more rarely, the word 'reflection'. Yet these facts were well known in 1920 and stand, perhaps even today, in need of interpretation. The specific interpretation which Benjamin offers is unhelpful mainly for two reasons. First, as we have seen, it conflicts with considerable evidence in Schlegel's writing (Sections 2.2.2–3). Second, we cannot credit Benjamin with the achievement of understanding Schlegel better than did Schlegel himself; since Benjamin's 'theory' of *Reflexionsmedium* is obscure and admittedly paradoxical, it fails to clarify Schlegel's (paradoxical) pronouncements on art. Furthermore, Benjamin's plausible concession that the 'theory' of *Reflexionsmedium* is obscure and paradoxical is in conflict with any theoretical value that we may wish to attribute to the notion of the ego-less reflection of art *in* art.

However, Benjamin presents that notion not only as part of a 'theory' but also as deriving from a free choice. I quote again: 'One is free to choose the metaphysical quality which one wishes to ascribe to this indifference-point of reflection. [...] In an Early Romantic understanding, the centre of reflection is art, not the I.' Contrary to Benjamin's wording, there may be no possibility of attributing different metaphysical qualities to the indifference-point of reflection. If anywhere, the 'metaphysical quality' of the absolute seems to lie in its very absoluteness. What Benjamin must mean is that it is a matter of free choice in which realm we place the absolute, and especially, that it is a matter of free choice whether we place it in the realm of subjectivity or in that of art. Are we, then, free to put the absolute wherever we like?

From a Fichtean point of view, we are not. Fichte's notion of the absolute I, at least in the relevant years between 1794 and 1797, is not part of a philosophy which aims at a theory of the absolute. His aim is to offer a post-Kantian, transcendental explanation of consciousness, and the 'absolute I' is a by-product of this attempt which reflects the predominance of the subject in the transcendental tradition. If we – together with Benjamin, the later Idealists and, at times, the Romantics – are unhappy with the epistemological predominance of the subject, we cannot choose a new name for what in Fichte is the absolute I without creating confusion. We *are*, of course, free to develop a version of Idealism that credits something else with the absolute status that in Fichte pertains to the I. As far as I can see, however, no existing version coincides with the 'theory' that Benjamin attributes to the Romantics. Put schematically, in Hegel we have the state, absolute knowledge and reflection, but no (unmediated) immediacy. The later, post-Fichtean, Schelling proposes a number of absolutes, including art; we also get a lot of immediacy, but no reflection. In Schlegel's Early Romantic poetological postulates, as Benjamin is fully aware, we do not possess any discursively developed philosophy at all. And according to Benjamin's own plausible assessment, the 'theory of *Reflexionsmedium*' is obscure and paradoxical. It does not qualify, then, as an intellectually attractive option.

The choice that Benjamin invokes is not an intellectual choice. In Part Two, he clearly states that all the concepts of art theory can be drawn into the realm of the absolute only at the cost of conceptual vagueness (98; *172*). And speaking of the Romantics, he clearly endorses the arationality which characterises the choice of the absolute of art. 'One would obviously be completely mistaken if one sought a particular reason why the Romantics considered art a medium of reflection. For them, this interpretation of all reality, hence also of art, was a metaphysical credo' (62; *149*). There is not only no (particular) reason to consider art a medium of reflection, but whoever as much as expects one is also wrong, and is so 'obviously' and 'completely'. The difference between a rational decision and that kind of choice which leads to the central idea of Benjamin's thesis, in consequence, must be very great. Indeed, choosing the absolute of art, in *The Concept of Art Criticism*, resembles that allegedly creative choice on which Benjamin rested his entreatment of the new God in 'The Religious Position' (GS II.1, 73 and Section 1.3.3). For him, the (Romantics') arational choice of the absolute of art must somehow mean that that absolute exists, for it is the examination *of that absolute* to which he explicitly devotes Part Two of his thesis (40; *134*).

In section 2.1 we saw that Benjamin's mentor Wyneken considered the experience of art especially suited to support the arational affirmation of religion. To a person like Benjamin, who encountered Wyneken's 'heroic' decisionism as an important part of his intellectual upbringing, it may indeed seem obvious that there cannot, and need not, be a particular reason for ascribing an absolute status to art. By the same token, however, Benjamin's endorsement of decisionism with regard to this central issue of his thesis means that ultimately *The Concept of Art Criticism* does not overcome the (intellectual) arbitrariness of Wyneken's religious and aesthetic postulates. Wyneken's 'fanatic pupil', who years after the break with his mentor continues to endorse religious decisionism, must have felt confirmed in this by his study of Schlegel. The subject's unrestricted authority in matters divine is an important theme in Schlegel's Early Romantic, experimental treatment of religion. In his note 90 (*note 93*), Benjamin refers to *Athenäum* fragment 234 (KA II, 203), in which Schlegel claims that perfect Christianity is Spinozism and that not solely Jesus but everything ought to have the status of a religious mediator. Even more outspoken is the following private note, dating from between 1796 and 1798: 'The belief that there is only one God is a blasphemous error. It is foolish to think that one ought to have only *One* mediator; for the true Christian, everything mediates. How many gods someone wants to have depends exclusively on his absolute volition' (KA XVIII, 56). The characterisation of monotheism as blasphemous, of all things, shows that the idea of orthodoxy and the related assumption of an authority which decides on the correct beliefs is not as absent from Schlegel's declaration of religious freedom as one might think. The authority in question is no longer found in traditional religious institutions but in the individual or rather, since 'blasphemy' does not cease to exist, in some individuals. Resting on absolute *Willkühr* ('volition'), the envisaged new religious authority resembles a pre-modern, absolute monarch. Its decisions are not necessarily devoid of any reason or irrational, but they are effective without a reason, and no possible reason, as a reason, possesses either a binding or a disqualifying force. By calling it 'obviously' and 'completely' mistaken to expect a reason for the Romantics' choice of the god *Kunst*, Benjamin may or may not overstate the amount of arbitrariness in Schlegel's poetics, but is certainly in line with the arationality of Schlegel's Early Romantic religious experiments. The congeniality between the two authors lies not in a shared epistemological or poetological conception of self-reflective literature but in their common interest in a new religion, their readiness to base such a religion on the subject's decisions, and the attempt to realise their own decision – sometimes and not exclusively – in the realm of poetics.

Both authors seem unaware of the philosophical comedy contained in their embracement of religious creativity. Schlegel writes that it is a matter of absolute volition how many gods someone *wants* to have; and Benjamin comes to his assistance by claiming that one is free to choose the metaphysical quality which one *wishes* ('möchte') to ascribe to the absolute. Both contentions are as certainly true (in modern societies) as their truth is inapt to provide the envisaged new religion with any credibility at all. 'Absolute monarchs' cut a funny figure on the stage of thought, however legitimate their existential desires (and potentially dangerous their effects). But there is an important difference between the two authors. As illustrated for instance by *Lyceum* fragment 54 (KA II, 154), Schlegel is enough of a satirist to perceive how grotesque the suggestions of someone can be who assumes for himself an unrestricted authority in religious matters. There are writers, he observes, who drink the absolute like water, and books in which even dogs relate to the infinite.[121] Dogs, then, make *no* trustworthy mediators after all. In Benjamin's deadly serious metaphysical phantasies there seems to be nothing that could match the wisdom behind Schlegel's satirical edge. The satirical insight of the Early Romantic writer, moreover, can help to explain his later attempt to make peace with traditional religion. For Benjamin, Schlegel's engagement with Catholicism must have amounted to the self-betrayal of an intellectual and opportunist, for he saw his own life-long refusal to comply with traditional religion as an expression of intellectual integrity and radicalism.[122] However, if not to betray one's intellect means to shape one's beliefs as much as possible in accordance with what can be known intellectually, there is, in the Schlegel of 1800, not much intellect to be betrayed in the first place. Nor does the value of intellectual understanding rate highly for the Benjamin of the 1910s. Once the intention to limit oneself to what can be known is gone, the acceptance of a traditional religion is a serious option apart from any social advancement it may bring. For in the last analysis, 'everything is self' is not only intellectually doubtful but also religiously counter-productive. Any sanctification must sanctify what is profane, yet no belief in sanctification can do without a borderline between the sacred and the non-sacred, without, to put it differently, unholy dogs of some description. The acceptance of a traditional religion does justice to the fact that I must not experience that borderline as my own creation if sanctification is to work. Whether or not one can choose one's gods, one can certainly not choose a sanctification achieved exclusively by oneself. For the subject to experience its life as sanctified, the belief in a power higher than that of the subject is needed. Traditional religious authority, whatever else it may be, is also a tangible, worldly expression of this necessity. If one desires a new religion as the expression of creative freedom, one will eventually find an old religion – perhaps one will settle, like Schlegel and Rosenzweig, for an unconventional version of a traditional religion – or one will find no religion at all. To go empty-handed, as does Benjamin, cannot count as intellectual radicalism if the value of intellectual understanding has already been sacrificed on the way.

2.2.5 Reflection as self-transcendence

Benjamin sees the major achievement of Early Romantic poetics in the establishment of 'work of art' as a basic theoretical category. In his presentation, 'work of art' is closely related to the concepts of reflection and artistic form. The Romantic theory of the work of art, he claims, is the theory of its form; and the form of the work is an expression of that reflection which constitutes the essence of the work (71–3; *154–6*). The 'ironisation of form', moreover, which consists in the voluntary destruction of the finite form of the work does not result from any activity on the part of the artist but is an objective feature of the work itself (83–4; *163*). By virtue of its self-destruction, the work displays an affinity to the Romantic concept of criticism or is even essential to it. Criticism transforms or introduces the finite work into that infinity of art which Benjamin considers a universal *continuum* of artistic forms, and to which he also refers by 'the idea of art', 'the absolute work of art' and 'the absolute art form' ('die absolute Kunstform').[123] However, the envisaged critical transformation differs from an arbitrary infringement only if it is essential for the work to transcend its own positive, limited form. Somehow, the reflection that constitutes the essence of the work of art must amount to the self-transcendence of that work.

In this section, I shall examine Benjamin's discussion of the self-transcendence of the work in three steps. 1. I shall show that this discussion presupposes, and thus fails to substantiate, the formalist understanding of 'reflection' that Benjamin attributes to the Romantics;

notably, Benjamin's treatment of self-transcendence to a large extent ignores Schlegel's interest in *particular* features of artistic form and activity. 2. I shall relate Benjamin's claim that the self-transcending reflection of the work is the opposite of ecstasy to Schlegel's treatment of self-relatedness and ecstasy before, during and after the *Athenäum* period, arguing that Benjamin's claim misconstrues Schlegel's position by concealing its mythico-theological implications and its closeness to the aesthetics of both Storm and Stress and neoclassicism. 3. I shall conclude with an assessment of Benjamin's own 'objective' account of artistic self-transcendence.

1. According to Benjamin, the Romantic belief in the absolute of art is not only a matter of decision but also one of interpretation, namely of the interpretation of 'form' and, by implication, 'reflection'. Fichte, he claims, has already determined the *Urschema* of all reflection, 'the thinking of thinking', as form; and while Fichte interpreted this form as I, Schlegel interpreted it around 1800 as aesthetic form (40; *134–5*). As shown in section 2.2.2, however, Benjamin has actually not established Fichte's determination of reflection as form. Neither is Benjamin's explanation of reflection as the transforming activity of reflecting upon a form convincingly derived from Fichte. As used by Benjamin, moreover, the expressions 'form of form' and 'form of thought' may be too unspecific to be fruitful for a theory of art.[124] Finally, we are not given any reason to assume the existence of one thing, 'form', of which both the Fichtean I and the Romantic conception of artistic form could count as interpretations. In Part One of the thesis, that is to say, the correlations between reflection, form of thought and artistic form which Benjamin upholds face the dangers of vagueness, equivocation and, indeed, non-existence.

Part Two does not significantly improve matters. Benjamin's introduction of the idea that reflection becomes evident in the form of the work rests on what he presents as his findings in Fichte. The crucial understanding of reflection as the essence of the work creeps in *en passant*, without receiving any philological or philosophical discussion. And the twofold characterisation of form as both the expression and the possibility of reflection remains as unexplained as Benjamin's ontological conclusion that form, by virtue of being the possibility of reflection, *a priori* provides a principle of existence ('Daseinsprinzip') on which the work rests (72–3; *155–6*).

Benjamin's *dicta* on form and reflection do not result from a documented analysis of Schlegel's texts. Let us take Benjamin's use of *Lyceum* fragment 37 (KA II, 151) as a first example. This famous fragment is characterised by a wealth of rhetorical, anthropological and psychological implications that seems to preclude any purely formalistic interpretation. Schlegel demands and justifies the artist's, and man's, self-restriction, yet also affirms the necessity of an enthusiasm which he associates with the idea of infinity. Self-restriction, he insists, presupposes the transcendence of the finite person; it is only possible where someone possesses an infinite power. The terms 'form' and 'reflection' do not figure in the fragment. In his comments, Benjamin *starts* from the notion that the self-restriction of reflection is form (74; *157*). It is only after this preparation that he quotes Schlegel's praise of both artistic and human self-restriction and claims that this praise concerns form as the self-restriction of reflection. Since the self-restriction *of reflection* is not explicitly discussed in the fragment, however, any substantiation of Benjamin's claim would require an analysis of the fragment. Such an analysis is not even attempted.

My second example shows that Benjamin takes the near-identity of reflection and form for granted, independent of the relation, in Schlegel, between 'self-reference' and specific artistic forms. It is only indirectly that Benjamin comes to consider the latter. In his discussion of irony he refers to Schlegel's appreciation of Greek Old Comedy and its *parabasis* – that is that part of the comedy in which the chorus advances to the proscenium and addresses the audience in the playwright's name – in order to illustrate the Romantic conception of the work of art such as it informs, for instance, Ludwig Tieck's comedies and Jean Paul's novels.[125] In his 1794 article on comedy, Schlegel defends Aristophanian *parabasis* as follows:

> This offence [against illusion and unity] comes not from a lack of skill but from level-headed wantonness, an exuberant plenitude of life; and it often has no ill effect at all but rather increases the impact, for it cannot really annihilate the illusion. The highest agility of life must act, must destroy; if it does not find anything external, it returns upon a loved object, upon itself, upon its own work; it then offends in order to stimulate, without causing destruction. [KA I, 30]

In using the phrase 'returns upon itself' ('wendet sich zurück auf sich selbst'), Schlegel evokes the literal sense of 'self-reflection'. Given the context of offence and destruction, moreover, the phrase comes close to suggesting a turning, of the agility of life, *against* itself. Schlegel uses the vocabulary of self-relatedness in this manner to argue for the fruitfulness of a specific artistic device, *parabasis*, against its neoclassical dismissal as destructive to the 'true' form, to the unity and illusion, of a play. The self-relatedness present in *parabasis*, he claims, is not really destructive but artistically fruitful; yet the highest agility of life which in *parabasis* (re)turns upon itself is also said necessarily to destroy. This means that Schlegel here brings together the issues of form, self-reference and a potentially destructive transformation which are essential to Benjamin's notion of form as a prefiguration, within the self-reflection of the work, of the destructive, or truly preserving, transformation of criticism (84, 86; *163*, *165*). Most surprisingly, however, when quoting from Schlegel's passage, Benjamin omits precisely those formulations by which Schlegel expresses the self-relatedness present in *parabasis* (84; *163*). By so doing, Benjamin deprives Schlegel's defence of *parabasis* of any relevance that it may possess for a theory of poetic reflection. Benjamin's own notion of literary self-reflection, to put it differently, is separated from Schlegel's concern with the specific formal features of Greek comedy, or, for that matter, Romantic (and romantic) irony. In *The Concept of Art Criticism*, Benjamin upholds the view of literary form as an expression of self-reflection irrespective of whether or not the vocabulary of self-relatedness is helpful for an understanding of those specific features of literary form with which Schlegel is concerned.

2. The omission in Benjamin's quotation from Schlegel's defence of *parabasis* has far-reaching consequences. The construction of *parabasis* as an instance of self-relatedness leads Schlegel to an understanding of 'poetic reflection' the wording of which, at the very least, is diametrically opposed to Benjamin's understanding of 'reflection' as the opposite of ecstasy (103–4; *175*). Poetic reflection, Schlegel writes in a note dating from 1806, 'is ecstasy' (KA XIX, 164). This note was still unpublished when Benjamin composed his thesis. As I shall now try to show, however, Schlegel's 1806 view of poetic reflection as ecstasy is linked, both genetically and systematically, to the 1794 construction of *parabasis* as an instance of self-relatedness, which Benjamin chose to ignore.

In his lectures on European literature of 1803–4, Schlegel gives an etymological explanation of *parabasis*; the term, he maintains, is derived from the Greek *ekbasis* ('stepping out of') as it was used for the movement of the chorus to the proscenium (KA XI, 88). In his notebooks, however, he uses the dramaturgical term to characterise irony and the novel. According to a philosophical note, irony is a permanent *parabasis*; according to a literary one, *parabasis* is a necessary ingredient in every novel (KA XVIII, 85, KA XVI, 265). *Parabasis*, which in the 1794 article was conceived in terms of self-relatedness, now serves as a metaphor for a self-relational 'stepping out of', for a 'stepping out of itself', of irony and the novel. Thus used, the term converges with Schlegel's understanding of 'reflection' as implying not only the return of something to itself but also the transcendence of its own boundaries.[126] Schlegel, moreover, links *parabasis* to 'Wollust' ('voluptuousness') and thus evokes the idea of ecstatic states (KA XVI, 334). This evocation also rests on the argument of the early article on comedy, especially on Schlegel's insistence on the Dionysian nature of comedy, the exuberant plenitude of life and the like, and related allusions to sexual love (KA I, 21–2). Since *parabasis*, in his use, is closely related to both 'reflection' and 'ecstasy', he can come to assert in 1806 that poetic reflection is ecstasy. By claiming reflection to be the opposite of ecstasy, Benjamin does not interpret Schlegel but dissociates himself from those vitalistic implications in Schlegel's proto-Nietzschean approach to Greek theatre that shape Schlegel's use of 'reflection'.[127]

Menninghaus defends Benjamin's claim as compatible with Schlegel's position. Benjamin, he argues, uses 'ecstasy' not like Schlegel as a dramatico-grammatical trope, but exclusively as the designation of that *mania* which according to Plato results from the poet's (and philosopher's) divine inspiration.[128] Schlegel's figurative uses of 'ecstasy' and *parabasis*, however, rest themselves on an idea of enthusiasm that is closely related to Platonic doctrine. Thus he presents his defence of *parabasis* as one which draws on the nature of comic enthusiasm and explains the genesis of Greek Comedy as an outpour of 'holy enthusiasm' (KA I, 30, 21). The 'highest agility of life', which in *parabasis* returns upon itself, moreover, coincides with 'holy enthusiasm' in counting as the source of comedy, for the (apparent) assault on unity and illusion can be considered a return of the agility upon its own work only if the comedy, or its form, is considered the work of the agility. Furthermore, the turning of the force upon the work can count as a *re*turn, of the force, upon itself only if one assumes that the generating force is still present in the generated work. In its construction of *parabasis* as self-relational, that is to say, Schlegel's apology for the Old Comedy rests on the Platonic assumption that the generating force is still effective in the work, an assumption with the help of which Plato explains the effect which the poet's song has on the audience.[129] Generally speaking, there can be no doubt that the early Schlegel both endorses Plato's 'magnetic' model of divine inspiration and links it to 'irony'.[130] The notes quoted above, in which Schlegel figuratively uses *parabasis* to characterise irony and the novel do not explicitly refer to the notion of the poet's holy enthusiasm. Yet in themselves they lack *any* context that would allow us to specify the meaning of Schlegel's figure of speech. It is therefore reasonable to draw for its explanation on the notion of *parabasis* in the 1794 article on comedy, which agrees with the Platonic model of inspiration. This is reasonable especially since in the context of his 1806 classification of poetic reflection as ecstasy, Schlegel again clearly presupposes the idea of divine inspiration: he dismisses as evil magic all poetic reflection or thought that is not accompanied by a reflection upon God (Notes 83–5, KA XIX, 164). Were 'ecstasy' merely a 'dramatico-grammatical trope', ecstasy could not count as evil. Only if we think of ecstasy as the poet's divinely caused self-transcendence can the failure to acknowledge God's work pass as a sinful presumption.

In the 1794 article on comedy, Schlegel 'naturalises' Plato's idea of an inspiring god by speaking of 'the force of life', 'the agility of life', and the like. Since at the same time he employs the vocabulary of holiness, however, such naturalisation does not go beyond the amalgamation, in the aesthetics of Storm and Stress, of divine inspiration with a natural, unconscious force at work in the genius.[131] Benjamin's assessment that the Schlegel of the *Athenäum* period overcomes the limitations of that aesthetics is therefore potentially misleading. It is misleading at least to the same extent that the implications of the early defence of *parabasis* are still present in Schlegel's poetological thought around 1800. The same holds of Benjamin's suggestion that in his 'theory' of reflection Schlegel overcomes the 'tendencies' of rationalism (70–1; *154–5*). In 1794, Schlegel does not reject the neoclassical criticism of Aristophanian *parabasis* comedy without implicitly affirming neoclassical norms at the same time.[132] Only by somehow endorsing the neoclassical conception of the work of art as a closed and illusionist whole, can he find, as he does, in Old Comedy a force that by offending against the rules of unity and illusion turns upon and against both the work and itself. A completely unclassically-minded author, in contrast, could not construe *parabasis* as an instance of self-reference; he could merely state the truism that there happen to exist literary works the authors (or generating forces) of which do not always comply with neoclassical rules. Arguably, the combination of Storm and Stress and neoclassical assumptions in Schlegel's early understanding of *parabasis* continues to characterise his poetological thought around 1800 to a very large extent. It is responsible for his Early Romantic attempts at reconciling enthusiasm and rationality in general, and his Early Romantic view of comical art in particular. The essence of that art, he states in *Athenäum* fragment 246, 'remains always the enthusiastic spirit and the classical form' (KA II, 206).

3. Although he declares reflection to be the opposite of ecstasy, Benjamin relies in his construction of the relation between the self-reflective work and the idea of art on the notion of self-transcendence, which Schlegel, the Greek language and the traditions of mysticism closely associate with 'ecstasy' (78; *159*). With the concept of formal irony, he attributes that self-transcendence which Plato locates in the inspired poet or philosopher to the work.[133] In support of the notion that artistic self-transcendence is an objective feature of the work for which the practice ('Verhalten') of the artist is irrelevant he quotes Schlegel's 'Herkules Musagetes' (84–5; *163*). In this poem, however, it is not the work but the artist to whom the advice is given to destroy the particular work in view of 'that work which no mortal will complete' (KA II, 418–19). Benjamin nevertheless opposes romantic or formal irony to subject-oriented conceptions of artistic self-transcendence and, while doing so, pretends to discard the notions of genius and substance in one go (86; *164–5*). Yet his recourse, in the same context, to 'the idea' and 'transcendental order' of art demonstrates his reliance on the Platonic theory of ideas, to which the notion of substance is essential. And given that formal irony is said to reveal the existence of the work 'as a mystery', the relation between the work and the idea cannot be the object of a profane and common knowledge. Benjamin's objective account of self-transcendence, that is to say, actually shares the esoteric implications of both the Platonic conception of inspiration and the cult of genius.

By conceiving the relation between work and idea in terms of 'absolute Abhängigkeit', Benjamin pays tribute to Friedrich Schleiermacher's famous grounding of religion in a feeling of absolute dependence.[134] This supports the view that Benjamin's attempt to escape the cult of genius remains tied to that subjectivity which is the burden of religion in modern times. His omission of Schleiermacher's term 'Gefühl' ('feeling') shows his struggle to shake this burden off. To the same extent, however, that the construct of the transcendence of the work as a non-ecstatic self-reflection is incomprehensible or, as Benjamin concedes, 'paradoxical', the striking difference which his dismissal of ecstasy seems to establish with respect to Wyneken's 'ecstatic' decisionism is devoid of theoretical substance (87; *165*). The description of irony as a 'storm' of revelation certainly removes it from any level-headedness. As regards the theory of the work of art in *The Concept of Art Criticism*, then, Benjamin's attempt to overcome the arbitrariness of his mentor's aesthetico-religious legacy is unsuccessful.

2.3 Criticism

2.3.1 The dilemma of 'immanent criticism'

Romantic criticism, Benjamin maintains, differs from formal irony by dissolving the form of the work 'irrevocably' ('unwiderruflich') and 'seriously' ('ernsthaft') (84; *163*). By calling the preservedness of the work in the idea eternal and indestructible, however, he also revokes the 'irrevocability' of critical dissolution (86; *165*). In the present section I shall discuss Benjamin's construction of his main subject, the Romantic concept of criticism, as one that shares the paradoxical character of 'formal irony'. 1. I shall present the author's lucid, though as such unacknowledged, recognition of what I call the dilemma of 'immanent criticism', the impossibility, that is, of ascertaining the self-transcendence of the work in the idea. 2. I shall discuss his attempt to limit the impact of this impossibility by loosening the connection between 'criticism' and the notions of *Urteil* ('judgement') and *Beurteilung* ('assessment'). 3. Finally, I shall examine his second attempt to limit the impact of the dilemma, namely his implicit suggestion that Schlegel's review of Goethe's *Wilhelm Meister's Apprenticeship* provides a solution.

1. The Romantics, Benjamin maintains, postulate that the work be criticised immanently, that is, that the standard ('Maßstab') for its criticism be provided by the work itself. It is clear that the work can only provide its own critical standard if it somehow differs from, or 'transcends', itself. Benjamin comments as follows: 'This postulate contains a peculiar paradox. For one cannot see how a work might be criticised on the ground of its own tendencies because these tendencies, in so far as they can be indisputably established, are fulfilled; and in so far as they are unfulfilled, they cannot be indisputably established.' He concludes that in 'the extreme case', this latter possibility means 'that inner tendencies are altogether lacking and that, as a result, immanent criticism would become impossible' (77; *159*).

The presence of a tendency in a work cannot be ascertained beyond possible doubt before the tendency is realised. If the tendency is realised, however, the work cannot be criticised for lagging behind the tendency. Either there is no critique or the existing criticism is not demonstrably immanent. At no point is it certain that there actually is immanent criticism. Benjamin does not clearly distinguish between this epistemic problem and the question whether immanent criticism is possible; and he tries to restrict the impossibility of the latter to 'the extreme case', which, incidentally, he considers elsewhere the suitable starting-point for theory.[135] If his epistemological argument is at all valid, however, it is so generally, and not only for the extreme case.

The impact of Benjamin's argument is devastating for his interpretation of Romantic 'criticism'. On the one hand, he must construe the relation between the work and the idea of art as the orientation or tendency *of the work* towards the idea. Without the idea of art being the work's own idea, the idea of art would be merely another extraneous 'rule' which the critic imposes onto the work, less specific, perhaps, than the rules of neoclassical poetics, but certainly just as 'heteronomous'. As a result, Benjamin's claim that Schlegel secures the autonomy of art by a theory of the work of art could hardly be more misleading. On the other hand, it is clear that the work can merely tend towards, but not be, the idea; as the universal and infinite *continuum* of all artistic forms, the idea of art is of necessity without a particular form. If form is a necessary element of any work, and if the theory of the work is indeed the theory of its form, the work necessarily differs from the idea. If we try to think of the work as realising the idea at the very moment of its self-destruction, there is again no work left which could be criticised. The epistemic dilemma which Benjamin identifies in 'immanent criticism' means that we cannot ascertain the existence of what the Romantic concept of criticism, as interpreted by Benjamin, denotes.[136]

2. The author tries hard to limit the damage caused by his sceptical insight. He presents the Romantic concept of criticism not as a version of the dilemma diagnosed but as its solution. The immanent tendency and the standard for the immanent criticism of the work, he claims, is the reflection on which the work rests. Before offering any explanation or argument in support of this claim, however, Benjamin changes its terms. In reality, he writes, 'this reflection is not so much the standard of assessment but, first and foremost, the basis of a completely different kind of criticism which does not aim at assessments, and which emphasises not the evaluation of the single work but the *Darstellung*[137] of its relations to all other works and, ultimately, to the idea of art'. (77–8; *159*) And he concludes that in its central intention, Romantic criticism is 'not assessment but, on the one hand, the perfection, completion and systematisation of the work and, on the other, its dissolution in the absolute. [...] these two processes coincide in the end.'

It is not clear whether Benjamin considers Romantic criticism as completely different from critical assessment or whether he sees it as striving primarily, but in addition to that assessment, for the perfection and dissolution of the work. This lack of clarity may echo Schlegel's remark, in 'Über Goethes Meister', that Goethe's novel should perhaps be assessed and not assessed at the same time (KA II, 133). However, Schlegel does not oppose the option of assessment to the self-relatedness of the work but on the contrary suggests that Goethe's novel assesses itself (KA II, 133–4). Benjamin's attempt to present 'assessment' as something opposed, or relatively unimportant, to the self-reflection of the work and the related notion of criticism is therefore questionable on textual grounds; it conforms perhaps more to his own inclusion, in 'On Language', of 'Urteil' ('judgement') among those mythic usages of language which mark man's distance from the absolute than to Schlegel's use of 'beurteilen' (GS II.1, 153; SW I, 71–2). More importantly, Benjamin's attempt does not provide what he needs. The dilemma of 'immanent criticism' applies to criticism as perfection no less than to criticism as assessment. If we cannot ascertain beyond possible doubt the inner, unrealised tendency of a work, neither can we ascertain beyond possible doubt that a particular piece of criticism is the perfection of the work being criticised, and not merely another work or text. This is so, irrespective of whether or not we call the supposed inner tendency reflection. In presenting the Romantic concept of criticism as the solution of the problem diagnosed in 'immanent criticism', Benjamin therefore confirms the belief, expressed already in 'On Language', that a paradox can serve as a solution.

The characterisation of the Romantics' critical intention as perfection, completion and systematisation corresponds to Schlegel's position in the supplement to the Lessing article. Schlegel there puts to hermeneutical use what he calls the 'organism of all sciences and arts' and conceives as possessing laws as well as a history (KA II, 411). Any individual work of art, he argues, can only be understood by determining its place and function in this world-historical organism of culture or *Bildung*. Benjamin's interpretation of 'the idea of art' as referring to the *continuum* of all artistic forms can be considered an (apparently) unhistorical version of Schlegel's world-historical *Bildung*; for as an organism, Schlegel's *Bildung* forms a *continuum*, and by virtue of its universality, the *continuum* which it forms is one of all (artistic) forms. We may, therefore, explain Schlegel's hermeneutical claim by saying that the understanding of a particular work requires the recognition of its intrinsic relation to the idea of art, and that for this reason criticism is the perfection, completion and systematisation of the work.

The textual support that can be found in Schlegel for Benjamin's characterisation of Romantic criticism, however, does not weaken the impact of the dilemma of 'immanent criticism'. Quite on the contrary, the laws and the history of Schlegel's world-historical organism of *Bildung* are the subject matter of a science that does not (as yet) exist (KA II, 411). Ignoring the laws and the history of the alleged organism, at least in their totality, we cannot rule out the possibility that world culture actually fails to form an organism. We cannot rule out the possibility, that is, that the relations of the particular work to all the others are merely external and that, as a result, the particular work lacks an inner tendency towards the idea of art. Nothing we know, no existing *Wissenschaft* entitles us to consider the critical dissolution of the work in the idea as different from a both fruitless and presumptuous exercise. What is needed to consider the dissolution – 'in the end' – perfection is a belief, for instance the belief in that 'divinatory criticism' which alone, according to *Atheäum* fragment 116, may dare to characterise the ideal of romantic, universal *Poesie* (KA II, 183).

3. Benjamin knows that the dilemma of 'immanent criticism' has not gone away, for he tries to counteract its impact in a second attempt, which, in some respects, is diametrically opposed to his first. Now dealing with Romantic criticism in the very terms of assessment and judgement which he previously discarded, he argues that 'the assessment of a work must never be explicit but must always be one which is implied in the fact of the romantic criticism (i.e. of the reflection of the work)'. A positive value judgement upon a work cannot be pronounced 'through a separate examination but solely through the very fact of criticism, for there is no other standard, no other criterion for the existence of a reflection than the possibility of its fruitful unfolding, which is called criticism' (78–9; *159–60*). In Benjamin's previous argument the reflection of the work counted as the standard of immanent criticism. Now it is the fact of immanent criticism that counts as the standard of reflection, indeed as the sole means which allows one to ascertain the existence of a self-reflection in the work.[138] The plausibility of 'immanent criticism' thus comes to depend exclusively on the availability of critical texts which demonstrate that, first, there is a self-reflection in the work under examination and, second, that they themselves succeed in criticising the work by unfolding that self-reflection.

Benjamin implicitly suggests that only Schlegel's Goethe review qualifies for these tasks. First, he comments only on the critical work of the Schlegel brothers, second, he criticises August Wilhelm's works as lacking in methodological reflection, and, third, among Friedrich's critical pieces it is only 'Über Goethes Meister' which for him corresponds completely to Friedrich's ideal of criticism by propounding at the same time a theory of criticism and a criticism of Goethe's novel.[139] The plausibility of the Romantic concept of criticism, then, seems to depend exclusively on the 'fact' of romantic criticism that is Schlegel's 'Über Goethes Meister'.

At this point, at the latest, an academic author would have had to reconsider the decision not to analyse the Romantics' critical practice, for there are not many important facts in literary matters which can be recognised directly, that is, without going through a process of analysis and interpretation (14; *118*). The view that 'Über Goethes Meister' demonstrates both the self-reflection of Goethe's novel and the critical unfolding of that self-reflection in the review itself is certainly not obviously true. Therefore Menninghaus, who holds with Benjamin that Schlegel's Goethe review corroborates the Romantic theory of reflection, rightly tries to support this assessment by an analysis of Schlegel's text.[140] However, far from corroborating the so-called Romantic theory of reflection, 'Über Goethes Meister' actually shares the problems of that 'theory'.

First, in roughly the first quarter of his review, Schlegel discusses the structure of *Apprenticeship* in so far as it affects the 'Gemüt' or soul of the reader (KA II, 126–30). If this analysis is relevant to Schlegel's notion that Goethe's novel assesses and *darstellt* itself (KA II, 133–4), it suggests that used in isolation, the vocabulary of self-reference represents a misleading abstraction from the process of communication between the work and the reader. While conforming to the general hermeneutical implications of Schlegelian thought, the view of 'literary self-reflection' as a crypto-hermeneutical construction is incompatible with Benjamin's work-oriented formalism in the thesis and elsewhere.[141]

Second, the reader whom Schlegel presupposes (or tries to create) is a recognisably Early Romantic one, congenial to Schlegel himself. For this reader possesses not only a 'sense of the universe', alias a 'presentiment of the whole world', alias a 'true systematic instinct' (KA II, 134), but also applies it to individual works of art (KA II, 131, incl. note 3). His desire to know how the whole is constructed is both, and at the same time, a cosmological and a poetological one. Schlegel's reader construction does not amount to a replacement of Goethe's novel with Schlegel's own thought[142] only if Schlegel is right in attributing to

Goethe's work 'the innate drive of the completely organised and organising work to form itself into a whole'.[143] However, the price which we have to pay for accepting this assessment, which contains the first important use of the vocabulary of self-reference in the review, is the non-identity of the work with itself: something completely organised *is* a whole in the relevant sense, and cannot *strive* for becoming one without differing from what it is.

Third, Schlegel links the recognition, and thus the successful self-characterisation, of the character of *Apprenticeship* to the presence, in the novel, of a spiritual nexus ('Zusammenhang') between the parts and the whole (KA II, 134). This means that he claims not only that Goethe's novel reflects – *darstellt*, assesses and can, if at all, only be understood from within – itself, but also that it does so only in part: 'Had this genius [of the book *Apprenticeship*] been capable of characterising itself in the whole as well as in the particular, no-one would be entitled to say more about what is actually in it, and how one has to take it. Here, it is possible to do a little more by way of completion' (KA II, 133–4). In its entirety, then, the novel is not its own, or its spirit's, unrestricted self-reflection. The self-reflection of *Apprenticeship* is at best an unfulfilled tendency. This result is fortunate for the Early Romantic critic: it saves his job. But it is detrimental to a theory of poetic self-reflection that relies on the practice of 'Über Goethes Meister', for it subjects such a theory to Benjamin's dilemma of the unascertainability of any (unfulfilled) inner tendency of a work. The dilemma of 'immanent criticism', which Benjamin tries to overcome with his appeal to the 'fact' of criticism, only reappears in the 'fact' – *the* Early Romantic fact of criticism, according to Benjamin and others – that is Schlegel's 'Über Goethes Meister'.

I conclude that Benjamin's second attempt at counteracting the dilemma resembles his first. In the second attempt, too, the 'paradox' is presented as a solution.

2.3.2 Mystical criticism

The incorporation, in Benjamin's thesis, of what he calls the paradox(es) of 'immanent criticism' demonstrates his awareness of the questionable epistemic value of both that concept and the related one of literary self-reflection. Taken in isolation, the epistemic problem identified by Benjamin may be less serious than its drastic description as a paradox suggests. The problem, after all, consists merely in the fact that we cannot be absolutely certain that there is self-reflection in a work and that its criticism can therefore be immanent. And, of course, it may not be advisable to expect literary theory, let alone criticism, to rely exclusively on assumptions the validity of which can be established beyond possible doubt. In conjunction with the indisputable and acknowledged paradoxes, however, which characterise the notion of self-reflection in Benjamin's presentation and elsewhere, the uncertainty diagnosed is indeed alarming. Apparently we cannot explain 'literary self-reflection' (without contradicting ourselves), *and* the sceptical view that we are unable to ascertain both the self-reflection of a work and the possible immanence of its criticism is justified. Given the semantic and logical problems of 'literary self-reflection', Benjamin's sceptical insight is a strong inducement to abandon the poetological paradigms which rely on that idea. In spite of the drastic way in which he voices his scepticism, however, Benjamin does not do this. Although he concedes the paradoxical nature of what he presents as the Romantics' theory of reflection, he supports the related concept of immanent criticism in his thesis and, indeed, relies on it in his later work.[144] We can explain this support for, and reliance on, a concept the epistemic validity of which is actually doubted by the particular nature of the criticism and the critical 'theory' that Benjamin practises in his thesis (and elsewhere).

A sceptical insight, like any other, does not categorically require one to adopt a certain course of action. The insight that we are unable to ascertain the immanent criticism and the paradoxically described self-reflection of a work is no exception. This insight is likely to lead to the abandonment of poetological paradigms of self-reflection only if one expects from poetics the intellectually coherent explanation of identifiable literary phenomena. If what one expects from poetics is the accommodation of a desire for religion, however, the insight does not weaken such paradigms but actually increases their attraction. In his thesis, 'immanent criticism' and the implied notion of the self-reflective work link that absolute which Benjamin tries to capture as *Reflexionsmedium* to literature and criticism by suggesting its presence in the individual work and its true, 'Romantic', criticism. Given, however, that Benjamin defends 'immanent criticism' only as unascertainable – given that he may even defend it *quia absurdum est* – the presence of the absolute is also removed from human understanding. Benjamin's sceptical account of 'immanent criticism' fulfils a religious function *in conjunction with* his support for 'immanent criticism'. The co-presence of scepticism and support, a stumbling block for non-religious interpretations of *The Concept of Art Criticism*, serves the purpose of post-traditional, religious edification perfectly well. In evoking the proximity *and* the remoteness of the absolute, it agrees with cultic practice and the requirements of sanctification (Section 1.3.3). And it respects the post-traditional character of a 'new' religion by searching for sanctification not in traditional religious cults but in particular forms of artistic and critical practice. Given Benjamin's explicit understanding of the self-reflective work as a revelation-and-mystery, the sceptical undermining of 'immanent criticism' is a necessary ingredient of, and not a reason against, his affirmation of this concept (86; *165*). As such, it is in line with his pronouncements on theory, truth and criticism both before and after the composition of *The Concept of Art Criticism*. In the previous section we already saw that his defence of 'immanent criticism' in the thesis agrees with the programmatic requirement of 'On Language' to present 'the paradox' as a solution. As 'paradoxes', the self-reflective work, its formal irony and its immanent criticism also promise that 'redemption' of the desire for mystery which Benjamin expounds in 'On Riddle and Mystery' (GS VI, 17–18; SW I, 267–8). Moreover, the sceptical undermining of 'immanent criticism' conforms to the conception of truth in the *Trauerspiel* study, according to which truth is the revelation of the mystery *as such* (GS I.1, 211). And it conforms, by ensuring the impossibility of disclosure ('Unenthüllbarkeit'), to the idea of criticism upheld in 'Goethe's Elective Affinities' (GS I.1, 195–6; SW I, 351).

The concluding image of *The Concept of Art Criticism* reaffirms, compresses and radicalises the double strategy of support and scepticism of Benjamin's previous treatment of 'immanent criticism'.

> Die Absolutierung des geschaffenen Werkes, das kritische Verfahren, war ihm [Friedrich Schlegel] das Höchste. Es läßt sich in einem Bild versinnlichen als die Erzeugung der Blendung im Werk. Diese Blendung – das nüchterne Licht – macht die Vielheit der Werke verlöschen. Es ist die Idee. [119; *185*]

> For Friedrich Schlegel, rendering the created work absolute, the procedure of criticism, was the highest. It can be illustrated in an image as the production of the dazzling light in the work. This dazzling – the sober light – extinguishes the plurality of the works. It is the idea.

What criticism must produce cannot exist beforehand. The idea does not exist in the work before criticism produces it. Its critical production, however, destroys the work together with the plurality of the works. At no point, that is to say, is the idea in the work. With his concluding image, Benjamin ontologicises the epistemic problem that he identified in 'immanent criticism'. It is no longer mainly the possibility of ascertaining the idea in the work which he denies, but also the very presence of the idea in the work. What he earlier envisaged only for the 'extreme case', namely that there are no inner tendencies in the work, is now implicitly presented as the norm. As a result, we cannot by right consider criticism, or the rendering absolute of the work, 'immanent'. By implicitly denying his explicit support for 'immanent criticism', Benjamin, in his way, may heed his sceptical insight. Yet he also, and at the same time, reaffirms 'immanent criticism' by conceiving the procedure of criticism as the production of the idea 'in the work'.[145]

The co-presence of affirmation and scepticism characterises not only the concluding image of Benjamin's book but also the final image of the thesis proper:

> Kritik ist die Darstellung des prosaischen Kerns in jedem Werk. Dabei ist der Begriff "Darstellung" im Sinne der Chemie verstanden, als die Erzeugung eines Stoffes durch einen bestimmten Prozeß, welchem andere unterworfen werden. So hat es Schlegel gemeint, wenn er vom Wilhelm Meister sagt, das Werk 'beurteilt sich nicht nur selbst, es stellt sich auch selbst dar'. [109; *178*]

> Criticism is the *Darstellung* of the prosaic core in every work. In this, the concept *Darstellung* is understood in the chemical sense of 'preparation', as the production of a substance through a certain process to which other substances are subjected. This is what Schlegel means when he says of *Wilhelm Meister* that the work 'not only assesses, but also *darstellt* itself'.

Prose ('Prosa'), as Benjamin has just maintained, can be called the idea of *Poesie* since it is the place where the reflective medium of poetic forms manifests itself (102–3; *174–5*). The *Darstellung* of the prosaic core in every work is therefore, at least in part, the *Darstellung* of the idea in the works.[146] The chemical model of preparation implies that the idea is absent from the works before criticism produces it. A substance that can be produced from other substances, after all, is not present in the latter; no substance *is* its own potentiality. The chemical model also implies that the critical production of the idea coincides with the destruction of the works: once the new substance is generated, the old ones have ceased to exist. Yet there is a difference between the model of preparation and the concluding image of Benjamin's book. While both figure in his treatment of Schlegel's concept of criticism, the model of preparation can be applied more easily to Benjamin's own approach in the thesis.

The model allows one to understand why the author of *The Concept of Art Criticism* is so unimpressed by what I called the methodological problems of his approach. In section 2.2.1, we saw that Benjamin is aware of the fact that the theory of absolute reflection that he presents as Schlegel's is actually absent from his sources. If we think of the thesis as a quasi-chemical attempt to 'prepare' the literary absolute from Schlegel's texts, that absence turns from a philological calamity into a familiar circumstance at the critical chemist's workplace. Of course, the substance that the chemist sets out to produce does not exist as yet. The chemical model of criticism (and philosophy), moreover, forms an actual part of Schlegel's Early Romantic writings.[147] Benjamin's attempted production of the literary absolute can therefore be described as the application of a Schlegelian model of criticism to those of Schlegel's writings around 1800 that contain 'criticism', 'reflection', '*Poesie*' and so forth. As a result, the so-called theory of *Reflektionsmedium* can be considered Schlegelian in kind. (It remains misleading, however, to present it as Schlegel's.) In so far as the critic's work resembles the chemist's preparation, finally, the critic need not be troubled by the paradoxes which he produces: the principle of non-contradiction does not apply to changes in matter. This means that the chemical model of criticism can in principle justify what independent of it are the absurd shortcomings of Benjamin's book. Actually, however, the model does provide such justification only if its basic magico-mystical assumption is correct, namely the assumption that linguistic phenomena – words, concepts, meanings, perhaps ideas – function as (living) matter. The experiment *The Concept of Art Criticism* is successful only if there is, on the one hand, an essentially linguistical *communicatio realis* (Section 1.1) and, on the other, an unfoldable 'germ' in human reflection, thought or art. As a mystic text, that is to say, the book and its 'method' make as much (non)sense as does linguistic mysticism itself. With his thesis, or the 'Tat' ('deed') of its demonstration, Benjamin at least in part strives for producing what he calls a 'fact' of criticism (41, 79; *135*, *160*). Yet just as in the case of Schlegel's 'Über Goethes Meister', the acceptance of *The Concept of Art Criticism* as such a fact presupposes, and does not represent a corroboration of, the implied concept of criticism and its basic mystical assumptions.

Benjamin characterises his subject matter and method as follows: '[...] the Romantic concept of criticism itself is an exemplary instance of mystical terminology, and for that reason the present study is not a reproduction of a Romantic theory of art criticism but the analysis of its concept' (50; *141–2*). Benjamin plausibly maintains that Schlegel's mystical terminology attempts to combine a minimum of discursive thinking with a maximum of systematic importance (47; *139*). The classification of Schlegel's 'concept' of criticism as mystical therefore means that there exists no (developed) theory of criticism in Schlegel in any discursive understanding of the word 'theory'. As a result, *The Concept of Art Criticism* cannot be, and is not, the reproduction of such a theory. This is Benjamin's negative characterisation of his book. According to his positive characterisation, the book is the 'analysis' of the Romantic concept of art criticism. Taken in isolation, 'analysis' is likely to suggest that Benjamin's purpose is a discursive investigation. However, the discursive analysis of a concept is the analysis, focusing on that concept, of the theory in which the concept figures. And as we have just seen, Benjamin's view of Schlegel's 'criticism' as a mystical term rules out the existence of such a theory in Schlegel. A little earlier, moreover, Benjamin stated that with his 'analysis' of the concept of criticism he aimed at demonstrating its philosophical fruitfulness for the determination of the reflective medium as art (44; *138*). Given that he considers the 'theory' of *Reflexionsmedium* self-contradictory, this means that his analysis of 'criticism' is to demonstrate the philosophical fruitfulness of 'criticism' for a certain interpretation, or application, of a paradoxical theory. What Benjamin calls his 'analysis' of the Romantic concept of criticism, then, is of necessity in conflict with any way of thinking which adheres to the principle of non-contradiction. His 'analysis', that is to say, does actually not promise a discursive inquiry but agrees with the notion of analysis in Schlegel's *Athenäum* fragment 83: 'The principle of non-contradiction is not even the principle of analysis, namely of absolute analysis, which alone deserves this name, of the chemical decomposition of an individual into its downright elementary components' (KA II, 178).

Athenäum fragment 83 is part of Schlegel's 'chemical' conception of criticism, with the evocation of which Benjamin finishes his thesis. In *Athenäum* fragment 242, Schlegel elaborates on his paradoxical notion of individual by suggesting that all individuals – *even those which are individuals in the most literal sense* – are complex entities: 'Are not [...] all individuals systems, at least in their germ and according to tendency?' (KA II, 205). Benjamin draws on the idea of a systematical 'germ' to characterise Schlegel's mystical terminology for which he takes, as we have seen, the 'concept' of criticism to be of exemplary importance:

'Terminology is the sphere in which, beyond discursivity [...] his [Schlegel's] thinking moves. For the *terminus*, the concept, for him contained the germ of the system, was basically nothing other than a prefigured system itself' (47; *140*). According to this description of Schlegel's thought, concepts contain the germ of a system and thus do what individuals do according to *Athenäum* fragment 242. Benjamin's description finds support in Schlegel since *Athenäum* fragments 83 and 242 deal under the heading 'individual' with linguistic or language related phenomena,[148] and Benjamin's quotations from the Cologne lectures demonstrate Schlegel's corresponding use of 'concept'. However, if a single concept prefigures, or contains the germ of, 'the' system, in his analysis of that concept the analyst can discover the scheme of the all-encompassing system or, for that matter, the absolute.

If a single term can contain a system, moreover, the term 'Reflexion' in *Athenäum* fragment 116 can in principle contain, as Benjamin suggests, the 'theory' of *Reflexionsmedium* (63; *150*). The term 'Reflexion' does so indeed if two further requirements are fulfilled. It must be possible to relate those components of the 'theory' that are not directly derivable from *Athenäum* fragment 116 to other statements by Schlegel (or the Romantics). And all the terms used must be part of a *continuum* of concepts. Benjamin accordingly maintains that the existence of such a *continuum* is a presupposition of mystical terminology: '[Mystical terminology] is the attempt to call the system by its name, that is to say, to comprehend it in a single mystical concept in such a way that the systematic interconnections are contained in this concept. This presupposes that the concepts form a continuous medial connection, a medium of reflection' (48–9; *140*). This passage strongly suggests that Benjamin shares Schlegel's mystical intentions. First, the original of its last sentence can also be interpreted to the effect that the presupposition of a *Reflexionsmedium* of concepts is actually met ('liegt [...] vor'). Second, the colloquial expression 'etwas beim Namen nennen' ('to call something by its (proper) name') is superfluous for the characterisation of Schlegel's mystical thought *and* stands out from the predominantly formal style of the thesis. Interpreted as a pointer to Benjamin's own mystical thought, however, the idiomatic expression is informative and no longer colloquial. As shown in section 1.2.3, the author of 'On Language' attempts to incorporate a would-be metaphysical 'system' in one term: 'name'. Like Schlegel, that is to say, Benjamin aims at combining a maximum of systematical importance with a minimum of discursive argument. Since 'On Language' was published only posthumously, most of the original readers of *The Concept of Art Criticism* of necessity ignored the fact that by 'name' Benjamin links Schlegel's mysticism to his own writing. This does not hold, however, for the other indications of mysticism in the book. In the passage quoted, Benjamin uses his very own term 'Reflexionsmedium' to characterise a basic presupposition of Schlegel's mystical terminology. More generally, he gives an account of his approach in *The Concept of Art Criticism* that closely resembles that given of Schlegel's mysticism. His own study is distinguished from any 'purely systematic' one and Charlotte Pingoud quoted to the effect that Schlegel could not limit himself to the 'purely systematic'. *The Concept of Art Criticism* counts as 'oriented towards a system', and Schlegel's mysticism as relatable to systematic thought and full of systematic intentions (12, 41–5; *117, 137–8*). Finally, Benjamin's treatment of Schlegel's mystical style of thought in the first, 'systematic', part of the study is not misplaced only if Schlegel's mysticism is relevant to Benjamin's own 'systematic' ambitions.[149]

Benjamin's characterisation of his method as 'analysis', then, is best understood as a declaration of mystical intent which agrees with Schlegel's 'terminological' mysticism, as discussed in the third chapter of Part One of *The Concept of Art Criticism*. Such agreement appears to be ruled out by Benjamin's classification of Schlegel's mysticism as 'verhängnisvoll' ('fateful', 'disastrous') (45; *138*). In reality, however, it is not. The word 'verhängnisvoll' surrounds Schlegel's mysticism with an aura of destiny and thus presents a use of language as sublime which otherwise may look merely confused. And as regards the possibility of disaster and catastrophe which 'verhängnisvoll' evokes, we know Benjamin to be not exactly faint-hearted. In conclusion, let me return to the final image of *The Concept of Art Criticism*.

In the name of the idea, this image celebrates as sacred the destruction of human sight and insight, for no 'sober' light which is not at the same time holy will ever blind dazzlingly.[150] The image, moreover, obliges criticism to function as the literary counterpart of that methodological nihilism which, according to the theologico-political fragment, fulfils the task of world politics by striving for the messianic transience of both nature and history (GS II.1, 203–4). 'Messianic art criticism,' as Benjamin's friend Florens Christian Rang puts it, 'is but a special instance of messianic world critique, of Judgement.'[151] To the critical extinction of the finite works of human culture which Benjamin's image depicts corresponds the end of the world that the fragment evokes under the headings of both *Untergang* ('fall', 'destruction', 'apocalypse') and happiness. In epistemic regards, the political nihilist is as unsheltered as is the Benjaminian critic of literature. There is no demonstrable relation between the historical and the messianic, just as there is no ascertainable immanence of the idea in the work. The implicit ontologicisation, in the image, of the epistemic dilemma of 'immanent criticism' is matched by the claim in the fragment that it is only the Messiah who will create that relation. However, Benjamin's fragment as does his image also upholds the idea of immanence. The Messiah is said not only to create the (non-existent) relation between the historical and the messianic, but also to redeem and to complete it. Therefore, this relation must also somehow precede the Messiah's advent. The religious meaning of history seems to exist as a tendency and a suffering which find completion and redemption respectively at the end of times. By his promotion of *Untergang* and suffering, the political nihilist urges messianic fulfilment. He shares the hope, expressed already in 'The Religious Position', that the painful intensification of the desire for the sacred will bring revelation. In his preparedness to sacrifice all human concerns, he strives for that religious heroism which in *Fear and Trembling* Kierkegaard admires in Abraham's attempted infanticide, with the difference, of course, that the nihilist is a 'knight' not of faith but of desire. The epistemic insecurity of his strategy means that the risks that he takes are high. If the Messiah who both completes and transforms his courtship of disaster fails to come, the nihilist will only achieve *Untergang* but no happiness. Fortunately, the risks implied in 'messianic art criticism' are considerably smaller. In pursuing the paradoxical task identified by Benjamin in romantic 'criticism', the critic will not really destroy any work. He will only produce pieces of criticism that are intellectually unconvincing and, perhaps, as sad for some as they are edifying for others.

3
History and *Technik*

3.1 Performative continuity

Although there is widespread consensus today that some kind of continuity exists between Benjamin's early 'metaphysical' works and his 'materialist' writing from the mid-1920s onwards, it is not clear how this continuity ought to be specified.[1] We cannot specify it in terms of theoretical content. Benjamin's programmatic pursuit of paradox with regard to those postulates of self-reference that underlie the early work makes it impossible to describe that work adequately by the extraction of a consistent set of central ideas. The early work fails *systematically* to provide a consistent theory of its apparent object domains, such as language and literature. (In contrast, it may successfully express the impossibility of a consistent first-level theory wherever a dominant desire for sanctification is involved.) This means that at least the earlier of the two poles between which the problematic continuity of Benjamin's *œuvre* unfolds cannot be defined in terms of theoretical content. And since the more penetrating reconstructions of his 'materialism' in the literature invariably seem to imply an understanding of 'materialism' that rests on, or is in agreement with, magical, idealistic or religious assumptions,[2] the same is likely to hold for the second pole, too.

While the systematic affirmation of central paradoxes undermines the formation of theoretical insight, it can promote a number of non-theoretical objectives. A pragmatic continuity in the performative dimension of Benjamin's *œuvre* may thus well derive from the very erection of paradoxes that thwarts the epistemic specification of continuity. In Chapters 1 and 2, I argued that the early work is adequately described as a mystic act or a post-traditional religious exercise. In Chapter 3, I shall show that this description remains valid for essential features of the later work.[3] In sections 3.2.1–2, I shall argue that the Benjamin of the *Trauerspiel* study implicitly applies 'the theory of *Reflexionsmedium*' to history, and that it is precisely his religious radicalism which explains the apparent dismissal, in that study, of romantic 'criticism'. Sections 3.3.1–2 will interpret Benjamin's advocacy of 'political materialism' or communism as a subsidiary rhetorical strategy which permits him to uphold, in the face of an increasingly threatening political religion on the Right, those religious intentions which inform his 'anthropological materialism'. The analyses in sections 3.4.1–3 will try to demonstrate the persistent importance of Benjamin's religious rhetoric of self-reference in his understanding of, respectively, 'technique' and 'technology', Brechtian theatre, and politics and film. In conclusion, I shall comment on his last important text, 'On the Concept of History', in the light of the argument developed in the present study.

3.2 Mortification

3.2.1 Historical self-reflection

In the *Trauerspiel* study, Benjamin opposes his understanding of 'criticism' to a romantic one: 'Criticism is the mortification of the works. [...] Mortification of the works: not then – as in romantic thought – the rousing of consciousness in living works, but the settlement of knowledge in works which have withered' (GS I.1, 357). This definition of 'criticism' seems to imply a rejection of the idea of literary self-reflection as developed in Benjamin's thesis.[4] Romantic thought, Benjamin argues, conceives the work of art as a living and conscious entity, and criticism as that which activates the consciousness in the work. This consciousness – as indicated by Benjamin's reference to *The Concept of Art Criticism* in his note 55 – is conceived as reflection and the arousal of it by criticism, accordingly, as the development of the work's own reflection. Criticism as mortification, in contrast, does not rouse the self-knowledge in living works but places knowledge into works which, since they are 'abgestorben' ('withered', 'dead'), seem to possess neither a life nor any knowledge or self-consciousness of their own. By 'settling' knowledge in works to which knowledge appears external, criticism as mortification seems to run counter to the postulate of immanent criticism.

However, the very definition of criticism as mortification subverts the opposition between living and dead works on which Benjamin draws.

Works which can 'absterben' or 'wither' must be alive. More importantly, their life must somehow persist during their critical mortification, for one cannot render dead or 'mortificate' what is dead already. The understanding of criticism as mortification of course implies that the critical process would find its completion in the death of the work. Yet this implication, far from distinguishing Benjamin's understanding of 'criticism' in 1925 from his romantic one of 1919–20, is an essential aspect of the latter. In his thesis, Benjamin construed 'criticism' as destroying the form of the work and explicitly linked such destruction to Schlegel's notion, in 'Herkules Musagetes', that the artist ought to put his work to death (GS I.1, 84–5; SW I, 163). And the image for romantic annihilation with which *The Concept of Art Criticism* concludes finds a counterpart in the simile for the revelation of truth in art, and thus also for critical mortification, which Benjamin offers in the Prologue to the *Trauerspiel* study. The burning of the work in which the form reaches its highest luminosity corresponds to the dazzling extinction of the (plurality of the) works in the idea.[5] The question thus arises whether there is a substantial difference between Benjamin's conception of Romantic criticism and his notion of critical mortification.

Schwarz Wentzer's suggestion that such a difference lies in Benjamin's rejection of the Schlegelian equation of the work with the absolute is unconvincing.[6] On the one hand, there is no such equation in Schlegel which leaves the empirical concept of the work of art intact; where Schlegel uses the term 'work' with reference to the idea (of art) or the absolute, his use of it, as Benjamin argues in his thesis, is clearly of a mystical nature. On the other hand, in the *Trauerspiel* study Benjamin himself uses a mystical notion of the work of art. Employing the terminology of 'Goethe's Elective Affinities', he maintains that the transformation of the material contents into truth content grounds a rebirth on the decay of the *Wirkung* ('effectiveness', 'attraction') of the works. 'In that rebirth all ephemeral beauty completely dissolves and *the work asserts itself* as a ruin. In the allegorical construction of the Baroque *Trauerspiel*, such fragmentary forms of *the salvaged work of art* have always clearly stood out' (GS I.1, 358, my italics). In entering the realm of truth, the empirical work is burnt, but the highest luminosity of its form must amount to the mystical life of the work in the circle of ideas. It is only the fragmented or 'mortificated' work that is salvaged, but if it is indeed salvaged, idea and work are no absolute opposites, just as, according to Benjamin's thesis, Schlegel implies when considering the idea (of art) a work.

Nor does criticism as mortification differ from Benjamin's conception of Romantic criticism through the rejection of the postulate of immanent criticism.[7] In his thesis, Benjamin considered that postulate the (paradoxical) requirement that the work be criticised on the basis of its own tendencies. In the *Trauerspiel* study, he represents works of art as possessing a tendency that is reinforced or realised in their critical mortification. 'They are designed from the very start for that critical decomposition to which the course of time subjected them. [...] Criticism is the mortification of the works' (GS I.1, 357). It is the task of philosophical critique to show 'that the function of the art form ["Kunstform"] lies precisely in the transformation of such material and historical contents as underlie each significant work into philosophical truth contents' (GS I.1, 358).

These statements differ from what is positively developed in *The Concept of Art Criticism* by conceiving history as an agency or dimension of critique; the decomposition of the work of art in the course of time is classified as 'critical'. I shall return to this point. What matters at present is that for Benjamin this historical dimension of critique is clearly compatible with the structure of 'immanent criticism'. It is essential to the Baroque works to tend towards their decomposition in the course of time and it is this tendency, as Benjamin confirms in the last paragraph of the study, which makes their idea decipherable. Precisely because they are not produced to last forever in the phenomenal world, they can escape transience; they can be 'ursprünglich', that is, in Benjamin's sense of the term (GS I.1, 226). The incompleteness of the Romantic fragment is meant to point towards completion. Benjamin finds the same intention in the melancholic allegories of the seventeenth century. In these allegories, he maintains, transience is not so much signified as itself signifying, and is indeed offered as the allegory of resurrection (GS I.1, 405–6). This orientation towards another life is the tendency 'within' the Baroque work by which it qualifies for 'immanent criticism'.

The tendency becomes apparent 'in the curious detail of the allegorical references', which is what endures of the German *Trauerspiel* and which, as Benjamin insists, 'lives' through 'the structure' (GS I.1, 357–8). This insistence on the life of the detail through the structure captures the gist of the Romantics' organic conception of the work of art. With the allegorical signification of its enduring detail, the individual work of the Baroque partakes of an organisation in which the part lives through the whole. The critical transformation of material into truth content is still, though of course not exclusively, a function of the form of the particular work. It is this conception that underlies

Benjamin's understanding of the process of historical decay from – and '*in*' – which philosophical critique is to 'salvage' the work. Criticism as mortification amounts to the application of 'immanent criticism' to non-classical works of art and their post-history ('Fortleben'). Accordingly, in 'The Task of the Translator' Benjamin convincingly characterises the concern with the post-history of the works as a Romantic one (GS IV, 14–15; SW I, 257–8).

Steiner sees in the notion of mortification the rejection not of the principle of immanent criticism, but of the view that that principle finds its substantiation in the 'theory' of *Reflexionsmedium*, which Benjamin developed in his thesis.[8] However, although the Benjamin of the *Trauerspiel* study no longer uses the expression *Reflexionsmedium* to indicate his concern with the absolute, he clearly reaffirms the underlying idea and structure which characterised his former use of it. Both the idea of an intensive infinity and the structure of self-referentiality figure prominently in the *Trauerspiel* study. For the introduction of 'intensive infinity', Benjamin in accordance with his seventeenth-century subject matter resorts no longer to the Romantics, as he did in the thesis, but to Leibniz. The being which enters with its post-history the idea – for instance the being of the art work, or even that of its enduring detail which lives through the structure – is said to contain in its own idea the condensed and indistinct image of all the other ideas (GS I.1, 228). As in Leibniz's monad, all phenomena are represented in that image, and all truth must be contained in it since Benjamin conceives the representation in the image as the *objective* interpretation of the world. By his contemplative immersion ('Versenkung') into everything real, by his immersion even into the apparently isolated detail of a long-forgotten and 'dead' *Trauerspiel*, the critic strives to capture that intensive infinity which, in *The Concept of Art Criticism*, was attributed to reflection. The truth, moreover, which is revealed in the incorporation of the being into the idea, has a self-referential structure. This follows from Benjamin's earlier distinction, in the Prologue, between truth and a non-philosophical, or scientific, 'cognition' to which the quality of possession ('Besitzcharakter') is essential.

> Diesem Besitztum [der Erkenntnis] ist Darstellung sekundär. Es existiert nicht bereits als ein Sich-Darstellendes. Gerade dies aber gilt von der Wahrheit. Methode, für die Erkenntnis ein Weg, den Gegenstand [...] zu gewinnen, ist für die Wahrheit Darstellung ihrer selbst und daher als Form mit ihr gegeben. [GS I.1, 209]

> To this possession [of cognition], *Darstellung* is secondary. It does not already exist as something which *darstellt* itself. Precisely this, however, holds for truth. Method, which for cognition is a way of acquiring the object, is for truth the *Darstellung* of itself, and therefore given with it as form.

Truth exists as self-referential *Darstellung*, which, however, pertains not only to truth but also to what Benjamin calls method. 'Truth' and truth-oriented 'method' even tend to merge. The phrase '*Darstellung* of itself' in the last sentence may take up the idea of the self-referential *Darstellung* of truth, which Benjamin affirms in the last but one. Yet the complement 'of itself' relates syntactically to 'method'. Similarly, that with which method is given as form may still be truth, but the sentence after the passage quoted opposes the form in question not to cognition but to the *method* of cognition. In substantive, 'philosophical' insight, Benjamin argues, method is not a way of acquiring knowledge about something other than the method itself but the self-*Darstellung* of method (as well as of truth). Implicitly, he thus reasserts the idea of the 'self-cognition of a method, of something formal', which in his thesis was central to both the alleged immediacy of reflection and the notion of the absolute subject itself (Section 2.2.2). Accordingly, in the Prologue he discusses method and truth as form, and continues by emphasising the essential immediacy in truth (GS I.1, 210). This incorporation, into the *Trauerspiel* study, of the formalist, neo-Kantian features of Benjamin's earlier notion of reflection may strike the reader as incompatible with the ontological gravity of the Prologue, in which the idea counts as a given ('vorgegeben') being ('Sein') (GS I.1, 210). However, in his thesis Benjamin already drew on neo-Kantian formalism only to dismiss it eventually as insufficient for the reconstruction of Romantic 'reflection'.

Finally, even the determination of the medium of reflection as art, which figures so prominently in the thesis, is not negated in the *Trauerspiel* study. On the one hand, we have already seen that Benjamin's definition of criticism as mortification implies that the critical transformation of material into truth content is a function of *Kunstform* (GS I.1, 358). On the other hand, in the thesis art did not count as the exclusive realm in which the 'theory' of *Reflexionsmedium* was applicable. As long as its character as a medium of reflection was preserved, Benjamin maintained in 1920, the absolute could be interpreted not only in terms of art but also, for example, in terms of history (GS I.1, 44; SW I, 138). With the emphasis, in the *Trauerspiel* study, on historical decay as an agency or prefiguration of redemptive critique he does not therefore abandon the notion of '*Reflexionsmedium*' but only complements that art-centred interpretation of it on which he focused in the thesis. His concern with the post-history or 'Fortleben' of the works indeed amounts to the combination of two interpretations of the absolute, art and history, both of which he had in principle acknowledged as early as 1916.[9] What is more, 'Reflexionsmedium', in the context of *The Concept of Art Criticism* as a whole, actually invites an interpretation of the absolute in terms of history. First, in his discussion of the immediacy of reflection, Benjamin used 'reflection' to denote a transformation in thought and a transition between different levels of consciousness (Section 2.2.2). He used 'reflection', that is, to denote a process and not a state. Second, he placed the critical dissolution into the absolute in a 'temporal' infinity that he characterised with reference to Romantic messianism as a qualitative, or full, medium ('eine mediale und qualitative [Unendlichkeit]') (GS I.1, 92; SW I, 168). And third, to the extent that Schlegel's Goethe review represents a crypto-hermeneutical account of the postulate of immanent criticism (Section 2.3.1), Benjamin's model of romantic criticism favours the replacement of the dominant formalism in his thesis with a conception of the work of art which acknowledges the importance of its reception and history.[10]

In the *Trauerspiel* study, then, Benjamin pursues further, and relies heavily on, what in his thesis he had treated as the Romantic concept of criticism. There is substantial agreement between the earlier conception of Romantic 'criticism' and his 1925 notion of critical mortification. Without doubt, however, in the *Trauerspiel* study Benjamin *presents* the latter notion as a negation of romantic thought: criticism is the mortification of the works, not as in romantic thought ... but ... . In the light of the persisting importance of the 'theory' of *Reflexionsmedium*, I suggest that the reason for this anti-romantic presentation of critical mortification is not a theoretical one but lies in a difference of attitude or *Gesinnung*.[11]

3.2.2 Manly romanticism

Benjamin presents as non-romantic what by his own standards is central to Romantic thought in order to distance his approach in the *Trauerspiel* study from a perception of Romanticism that is in conflict with the religious radicalism of his writing. His religious motives for this presentation find expression in the diatribe against the 'romantic aestheticians' at the beginning of the second part of the study (GS I.1, 336–7). This diatribe combines a Kierkegaardian distinction between the ethical and the aesthetic with the defence of the radical paradox in the true ('echt') theological notion of symbol. Benjamin defends that paradox against a profanation which he locates in an aesthetic tradition spreading from neo-Classicism via Romanticism into the twentieth century. It is not the notion of a symbolic presence of the divine to which he objects – a notion which he actually upheld in 'On Riddle and Mystery' and reaffirms in the Prologue (GS I.1, 216) – but the facile manner in which the adaptation of that notion in modern aesthetics tries to iron out the paradox. For him, the 'romantic' aestheticians fail to present *the paradox* as solution; they fail to conform to the ideal of 'theory', that is, which he had developed in 'On Language'. The apparent dismissal of Romantic 'criticism' in the *Trauerspiel* study thus agrees with his long-standing objections to what he perceived as the mawkish pantheism of his time. Just as in his 1912 dialogue on religiosity he found a desecration and the gilding of emotions in a religiosity that tried to do without a radical distinction between the sacred and the profane, he now sees in the aesthetic concern with the symbolic the noncommittalism and the cosy luxuries in which the aesthete's 'beautiful soul' is happy to indulge. The comfortable rituals of aestheticist subjectivism belittle the absurdity of the objective, true and religious understanding of 'symbol'; they try to celebrate life in a world which the 'methodological nihilist' must seek to destroy.

Benjamin responds in two ways to the dilemma that the radicalism of his religious desire is absent from the contemporary adoptions of those traditions of romantic and idealist thought on which he himself draws. First, in the *Trauerspiel* study as elsewhere, he tries to exempt *Early* Romanticism from the criticism of conformist profanation that he directs at modern aesthetics; the sentimental dimness of the cheap, non-paradoxical 'symbol', he writes, increased only after the end of Early Romanticism.[12] Second, he marks his difference from the lukewarm embracement of the absolute in neo-idealistic aesthetics by employing a rhetoric of masculinity.

The opposition between Early Romanticism on the one hand and neoclassicism, High Romanticism and neo-Romanticism on the other, however, does not solve Benjamin's dilemma. The more strictly such an opposition is conceived, the more questionable it becomes on historical grounds. More importantly, Benjamin's notion of critical mortification *as different from romantic thought* relates to his own reconstruction of Early Romanticism and not to any (non-Benjaminian) dimness that may have spread after the *Athenäum* period.

Benjamin's use of 'dialectical' belongs to both his strategic periodisation of cultural history and his rhetoric of masculinity. It is not the attempted reconciliation, in German Idealism, of classical immanence with Christian transcendence that he characterises as 'dialectical' but the reversals in Baroque allegory and 'apotheosis'. Severed from the aim of conceptual mediation and bound to his emphasis on the paradoxical, 'dialectical' comes to denote something that defies any non-mystical comprehension. For the mystic, cognition and volition frequently seem to merge. Benjamin accordingly relates 'dialectical' to a strengthening of the will. In their vulgar recuperation of 'symbol', the romantic aestheticians are said to display a lack of force ('Unkraft') and 'dialectical steeling' ('dialektische Stählung'). They fall short of the 'so to speak imperative attitude' with which the paradox is recognised in the true and theological concept of symbol. And the neoclassical, 'undialectical' apotheosis of existence, which paves the way for the subsequent desecrating mawkishness of romantic religiosity, deprives the ethical subject of the required masculine stature ('männliche Kontur'). Benjamin does not discuss the failure of Early Romanticism to realise the radical potential which he ascribes to it in relation to his own Romantic thought, but compensates for that failure by placing his own thought in an imaginary position of masculine strength.[13]

The religious requirements of Benjamin's writing, moreover, explain his controversial construction of the Baroque as a secular period.[14] The construction of the German *Trauerspiel* and the seventeenth century in general as devoid of religious thought and consolation (GS I.1, 253, 257 and *passim*) turns the Baroque into a stage on which the author can voice his own despair about the secular life of the twentieth century. In this, the *Trauerspiel* study conforms to the notion of literary historiography which Benjamin will expound in 'Literary History and Literary Research' (1931): The time of the works being studied serves as a medium in which to represent – or even 'prepare' – the historian's own time (GS III, 290; SW II, 464). The past becomes the medium in which the critic both expresses his dissatisfaction with the present and anticipates the desired, radical and apocalyptical change. Accordingly, the suspense of religious consolation is itself suspended at the end of Benjamin's study. In the schema of allegory, we eventually read, transience is offered as the allegory of resurrection: the allegorist awakens in God's world (GS I.1, 405–6). Yet the mystical author has proven his 'masculine stature' by confronting what he presented, all along before the happy ending, as the desolate and unrestricted immanence of the seventeenth century. Like other mystical writers before him,[15] he has descended 'into the empty abyss of evil' in order to find 'the assurance of infinity' (GS I.1, 404).

3.3 The red flag

3.3.1 Ambiguity: 1931

During the completion of the *Trauerspiel* study, Benjamin starts to align his work with Marxism.[16] He discusses this alignment with Scholem in their correspondence of spring 1931.[17] Both writers consider Benjamin's 'materialism' rhetorical in character. Scholem stresses the discrepancy between Benjamin's so-called true method and the one suggested by his materialist terminology and accounts for this discrepancy in terms of a potentially fatal self-deception on Benjamin's part (GB IV, 27–8, 36). Benjamin characterises his association with communism as a symbolic act and questions Scholem's right to object to this practice of political symbolism:

> But do you really want to bar me from raising over my little writing-shop located right in the middle of western districts [of Berlin] – quite simply because of the imperious need to distinguish myself from a neighbourhood that I cannot help accepting, and for a reason – do you want to bar me from hanging the red flag out of the window, lecturing me that it is no more than a little shred of cloth? [GB IV, 25]

What Benjamin defends here is neither the truth of communism nor even a simple truthfulness of his engagement with it but the symbolic dissociation of his writing from its 'bourgeois' surroundings in the wealthy western suburbs of Berlin.[18] McCole's discussion of Benjamin's work within the changing intellectual fields of Weimar culture shows that a major impulse for Benjamin's alignment with the Left lay in his growing need to distinguish his thought from the *in some respects* similar thinking of those conservative elites who supported, came to support, or were easily enlisted by, National Socialism.[19] His Marxist rhetoric, at least initially, functions precisely as a subsidiary device to safeguard his desire for religion against its political degradation in the 'neighbourhood' of conservative anti-modernism.[20] In the *Trauerspiel* study, the dissociation of his attitude from the 'bourgeois' one of the 'romantic aestheticians' figured within a highly idealistic framework. The motto of the book, moreover, appealed to the authority of Goethe and the romantic postulate of the unity of art and science, or knowledge, and thus presented the author's deviations from conventional literary scholarship as a genuine *methodological* continuation of the idealist tradition. Given his long-standing religious radicalism, such an idealistic and traditional rejection of the establishment's traditional idealism became unworkable after 1925. In the increasing political polarisation of the Weimar Republic, the existence of a materialist and communist alternative to idealism and 'bourgeois' culture represented an important point of view from which Benjamin's 'outside', which he defended as Wyneken's true legacy, looked very much like merely another 'inside'. Benjamin's position, that is, threatened to lose that radical appearance which he considered intrinsically valuable.[21] In response, he hoisted the 'red flag'.

Given the discrepancy between Marxist materialism and Benjamin's religious concerns, however, this symbolic dissociation of his public image from radical conservatism was bound to require a considerable amount of theoretically unconvincing adjustments. Scholem's insistence on an unambiguous and honest stance expresses both an understandable concern for the friend's intellectual integrity and the refusal to acknowledge the importance of that political landscape which Scholem himself had left behind by his emigration to Palestine. Benjamin emphasises the diminishing room for political manœuvre in 1931 Berlin: 'Here they [other methods to dissociate oneself unambiguously from the bourgeoisie] do not exist. Here there exists not even this method. For with a certain justification you could call what I call unambiguous the height of ambiguity. All right, I reach an extreme' (GB IV, 26).

Benjamin calls his symbolic endorsement of communism unambiguous *and* concedes that, with some truth, it can be considered the height of ambiguity. Although there is allegedly merely 'a certain' justification for considering his 'red flag' ambiguous, he categorically rules out the existence, in 1931 Germany, of any unambiguous dissociation from the bourgeoisie. He links his ambiguous account of his (freedom from) ambiguity to the extremity of his position. Neither the extremity of his position, however, nor the ambiguous adoption of ambiguity, is a recent phenomenon in Benjamin's life. Long before his alignment with Marxism, the co-presence of religious scepticism and an uncompromising, 'unambiguous', rejection of the 'bourgeois' contentment with secular life had led him to adopt a style of writing and thought for which the requirement of ambiguity was programmatic. Although 'On Language' both asserted and practised such programmatic ambiguity, Scholem opposes the allegedly solid and unequivocally clear insights of Benjamin's philosophy of language to the self-deceptive ambiguity of his 'materialist' writings. He moreover locates the former in an unbroken continuity of the most 'genuine' traditions and works with a disjunction between terminological presentation and 'true' method that is thoroughly at odds with the basic assumptions of Benjamin's linguistic mysticism.[22] Scholem, that is to say, misunderstands the very character of that 'metaphysics' of language to which he urges Benjamin to return.

The 'materialist' Benjamin breaks with the idealistic self-presentation but not with the basic 'methodological' and existential guidelines of his earlier writing. If the new appeals to materialism result in intellectual inconsistency, they result only in what is part and parcel of a sceptic's dominant striving for religious edification. Accordingly, for Benjamin intellectual inconsistency does not qualify as a reason to reject any strategy that favours the realisation of his religious intentions.[23] The symbolic endorsement of Marxism is without doubt suited to mark the difference between Benjamin's desire for apocalypse and the fascist, state-centred and nationalist project of a political religion. Since such endorsement is also consistent with the intellectual inconsistency of what Scholem considers to be Benjamin's true method, we obtain a correct description of Benjamin's writing for revelation if we replace 'materialist' with 'theological' in Scholem's account. According to Scholem, in his writing Benjamin is making out 'a materialist promissory note' that he is incapable of redeeming, incapable 'precisely because of what is most genuine and substantive' in his person (GB IV, 35). Indeed, given his 'genuine' desire for religion, Benjamin cannot hope to substantiate his appeals to Marx's historical materialism. Nor, however, can he hope to substantiate his implicit reliance on traditional theology in the face of his substantive religious scepticism. If there is self-deception in Benjamin's 'materialism', there is also self-deception in his 'theology' or 'metaphysics'. Yet Scholem's charge of self-deception presupposes the understanding of Benjamin's writing as theoretical. As a plea for revelation, however, that writing is based on the epistemically unsupported decision not to rest content with the unavailability of something sacred. If followed in the work of a sceptical twentieth-century intellectual, this decision is likely to create ambiguities and misunderstandings, but it is not – not necessarily and not in itself – self-deceptive or dishonest.[24] And precisely if religious fulfilment is the highest good, the moral problem of a strategic alliance with a largely anti-religious school of thought may be negligible. Benjamin accordingly can describe his rhetorical alignment with Marxism as an irreproachable measure to achieve something else: 'A castaway who drifts on a wreck endangers himself even more by climbing to the top of an already crumbling mast. But from there he has the chance to give a signal for his rescue' (GB IV, 26).

To whom is the signal sent? Benjamin does not say. The castaway does not know whether someone exists who will receive, let alone respond to, the signal. The signal cannot ultimately be addressed to the proletariat, however, for in Benjamin's simile communism is the already crumbling mast, not the *Rettung*, the rescue or redemption. Sending the signal may imply references and appeals to 'the proletariat', just as in 'The Religious Position' the attempt to prompt the unknown god's revelation implied references and appeals to 'the youth'. Yet while Benjamin presents his waving of the 'red flag' as a dangerous and reasonable act to increase a desperately small chance of rescue, he does not (and cannot reasonably) seek redemption in communism. Nor, however, is his 'red flag' necessarily a cheap shred of cloth. Its hoisting represents a rhetorical strategy in which the *theoretical* insubstantiality of symbolic action is entwined with the possibly non-existent truth of religion.

3.3.2 'Materialism'

Benjamin's endorsement of 'materialism' is more differentiated than the frequently loose use of the term in his correspondence suggests. In the surrealism essay (1929) and the related, only posthumously published, notes, he distinguishes between political and anthropological materialism on the one hand and metaphysical materialism on the other.[25] The central assumptions of metaphysical materialism that only material objects exist or that inanimate matter can serve as the basis of materialism are dismissed as insufficient. 'Political materialism' stands for the support of communism in the political arena, 'anthropological materialism' for the emphasis on man's bodily ('leibhaft') and physical life as a creature. This life exceeds that of the individual: 'The collective, too, has a body.' Benjamin embraces both political and anthropological materialism especially because between the two the destructive absorption or re-formation of the individual's inward nature can be accomplished. 'Anthropological materialism' is said to correspond to the experience of the surrealists, as well as to that expressed in the works of Hebel, Nietzsche, Rimbaud and Georg Büchner.

While the rejection of philosophical or 'mechanical' materialism is compatible with Marxian thought, the close relation which Benjamin construes between 'revolution' and 'the innermost world of the collective body' tends to replace the economic basis of Marx's historical materialism with a physiological one. By its physiological orientation, moreover, 'anthropological materialism' is also in conflict with Karl Korsch's critical Marxism, which tries to minimise economic determinism in favour of the consciousness and practice of the revolutionary subject.[26] Put drastically, Benjamin's 'proletariat' and its intellectual and literary allies are not rational agents but animals driven by their 'innervations'. They are, moreover, religiously educated animals. According to one of the surrealism notes, the 'innermost' world of their collective body has to be penetrated by the revolution as its most genuine meaning ('eigenster Sinn'), as the Indian and Jewish religions were penetrated by their morality. In the published essay, accordingly, traditional religion is not considered the opium of the people but, compared to opium and other drugs, the more rigorous 'school' ('Vorschule') of the 'true and creative overcoming' of religious by '*profane illumination*'. A 'profane' illumination for which religion provides an outstanding training ground, however, is at least congenial to religion; it is much more likely to inaugurate a new religion than a new non-religious society.[27] In his notes, Benjamin maintains that the difference between surrealist illumination and religious experience does not lie in the fact that the former is not revelation but that it is a revelation of events which become revelatory only retrospectively in collective remembrance. This, however, is not sufficient to consider the illumination in question profane. The orientation towards the past as the sphere of revelation is a feature of many established religions. While Benjamin at times seems engaged in an attempt to create a private religion, moreover, it is appropriate to conceive religion as a social, or 'collective', phenomenon. Finally, the fact that the events in question are not believed to have been revelatory at the time of their occurrence may well reflect the religious scepticism of the person desiring illumination, and not the non-religious nature of the desired illumination.

For Benjamin, 'profane' illumination permits the creation of the new *physis* of the bodily collective; it is a condition of the possibility to establish the new, literally understood, body political. In this function, 'profane' illumination finds a competitor *or* medium in technology ('Technik'); for Benjamin also claims that the new *physis* organises itself in technology. Technology is one of those productive forces by the development of which Marx explains the progress of society towards communism, in which religion, the cultural expression or effect of private property, will have vanished. Benjamin's reference to *Technik* can therefore appear to reduce the affinity of his anthropological materialism with religion. The same effect is achieved when, in his discussions of the First World War and fascism, he links the destructive character of technology in modern society to the degeneration of the liberating potential of technology in capitalist relations of production. The explicit adoption of Marxist themes, however, does not express Benjamin's interests as a writer. While he positively pursues the project of 'anthropological materialism', notably with his work on Proust and memory, he nowhere seems to engage in what could be called a detailed dialectical analysis of social institutions and economic life. Elements of Marx's economic dialectics are adopted without theoretical discussion. Indeed, they are adopted precisely where Benjamin's own notion of man's technological unification with nature tends to become undistinguishable from similar visions in radical conservatism.[28] The (apparent) dissociation of 'anthropological materialism' from post-traditional religion, that is to say, is an effect of Benjamin's 'political materialism' – or his 'red flag' rhetoric – and not an expression of the direct or primary intentions that inform his writing. This assessment is supported by two facts.

First, both the notion of mankind's collective body and a related concern with *Technik* figure in Benjamin's anthropological thought before his alignment with Marxism. I am referring to the 'Schemata on the Psycho-Physical Problem' that Benjamin conceived in the early 1920s in substantial agreement with Erich Unger's *Politik und Metaphysik*,[29] and in the context of his own earlier notes on psychology and perception. In both the earlier notes and the 'Schemata', indisputably religious intentions are coupled with a religious scepticism that finds expression in the phenomenological suspension of the question of truth.[30] Benjamin's later reception of Marx, as documented in the Marx file of the Arcades project, is not a naive attempt to understand Marx's concept of technology, or to adopt Hugo Fischer's 'anthropological' interpretation of that concept, but a process which to a large extent is guided by Benjamin's own already established anthropologico-religious interest in *Technik*.[31] In Benjamin's psycho-physical schemata, the basic assumption is that body ('Leib') and spirit ('Geist') differ merely as aspects ('Betrachtungsweisen') and not as objects. *Leib*, 'the function of historical presence in man', grows into the *Leib* of mankind since individuation, the principle of *Leib*, ranges higher than the individual nature of separate human bodies. Benjamin, that is to say, here conceives mankind along Romantic lines as an individual ('als Individualität').[32] Since the 'individual' mankind must be a collective one, however, the difference to the bodily collective of the surrealism essay is not a substantial one. Through *Technik*, moreover, both inanimate and animate nature is incorporated into the body of mankind. The technological creation of mankind's universalised body coincides with the eschatological simultaneous completion and death ('Untergang') of bodily life and is linked – in a manner reminiscent of the relation between the historical and the messianic in the theologico-political fragment – to the resurrection of man's *corporeal* ('körperlich') nature. In his corporeal nature, man is spiritual and belongs not to history but to God. In the surrealism essay, the religious implications of Benjamin's 'anthropological materialism' find expression in the conception of the collective body space as the messianic realisation of intensive infinity or 'the world of universal and integral *Aktualität*'.[33] They are also present in the cryptic inclusion of 'Einsamkeit' ('solitude') among the modes of 'profane' illumination (GS II.1, 308; SW I, 216). For 'Einsamkeit', according to Benjamin's early anthropological schemata, consists in man's consciousness of his 'immediate dependence upon God'. 'Profane' illumination, we may say, is acquired in an 'experience' which matches Schleiermacher's definition of religion.

Second, with his notion of *Technik* Benjamin incorporates into 'anthroplogical materialism' not an economic understanding of societal change but the structure of his religiously inspired postulates of self-reference. In the later work, self-reference is no longer directly and intensively dealt with as in 'On Language' or *The Concept of Art Criticism*. Nor does it figure implicitly, as in the *Trauerspiel* study, within an openly idealistic framework. But as I shall now try to show, Benjamin relies even in his most 'materialist' writings, 'The Author as Producer' and 'The Work of Art', on that rhetoric of self-reference which he had developed as a young man.

3.4 *Technik*

3.4.1 Tendency and *Technik*

In German in general, and in Benjamin in particular, the word 'Technik' is used to refer to both technology and artistic technique.[34] A Benjaminian theory of *Technik* might therefore be expected to clarify the relation between art and society's technological use of natural resources. However, in Benjamin's early work *Technik* figures within a conceptual framework which to a large extent is characterised by its disregard of the social sphere. 'On Language' lists *Technik* as a parallel to art and one of those realms in which language, the principle directed at the communication of spiritual contents, manifests itself (GS II.1, 140; SW I, 62). *Technik*, that is to say, is conceived as a realm in which there is communion or communication between things, between things and man, and between man and God, but hardly any interaction between people. The Romanticism thesis anticipates the insistence on literary technique in the later work by a highly formalist conception of literary self-reflection that minimises the hermeneutical, inter-subjective, implications of Schlegel's concept of criticism. In 'Critique of Violence', the conception of *Technik* as the most genuine realm of 'pure means' and 'language' rests on the assumption that direct social interaction is relatively unimportant in politics.[35] Benjamin's intensifying contact, during the 1920s, with those strands of avant-garde culture and European modernism which emphasise the technical character of art is well documented; it notably comprises his engagement with contemporary architecture, Paul Valéry's literary and Soviet political constructivism.[36] These influences do not, at least not immediately, result in a replacement of Benjamin's mystical understanding of *Kritik* with a 'completely secular' conception of art, as Witte puts it.[37] In the critical writings of the late 1920s and early 1930s, the demand for a more 'technical' – as well as sociological, egalitarian and 'materialistic' – interest in literature coexists with the endorsement of Schlegelian and 'magic' criticism, the support for 'genuine' theologians – Rosenzweig and Rang as opposed to Theodor Haecker – and reformulations of the adaptation of self-reference to (literary) history in the *Trauerspiel* study.[38] In 'False Critique' (1930–31), it is by and large the latter adaptation that Benjamin now characterises as the requirement of a well-conceived 'Marxism' and an as yet non-existent deductive aesthetic. Accordingly, he no longer attacks his academic opponents, as in the *Trauerspiel* study, for not being paradoxical enough but for being too 'idealistic'. They are criticised for turning literary research into the lay practice of a cult in which the 'eternal values' are celebrated according to a syncretic ritual (GS III, 285–6; SW II, 460–1). The specific formulation and the greater ferocity of this criticism reflect the increased political pressure on Benjamin's own position, but do not contribute to the substantiation of 'Technik' as a theoretical alternative to the attacked cult of *Geist*.[39] It seems to be Benjamin's aim in 'The Author as Producer' to establish this alternative. This text, originally designed as a lecture to be delivered at the *Institut pour l'étude du fascisme*, discusses topical questions of literary politics as they arose on the Left in the context of both the increasing influence of fascism in France and the attempts among exiled German writers and intellectuals to devise meaningful strategies against Nazi Germany.[40] Benjamin, however, characterises his direct objective as theoretical: He wants to demonstrate (*'beweisen'*), and *not* only decree ('dekretieren'), that the decisive criterion for the political function of literature lies not in its spiritual but in its technical qualities.[41]

Benjamin's central claim is that the correct political tendency of a work contains its literary tendency. In accordance with his 'political materialism', he presupposes that the correct political tendency lies in all those features of literary production which support the proletariat in the class struggle. The literary tendency amounts to the literary quality of the work, which consists in either a progress or a regress of literary *Technik*. 'Tendency', that is to say, is used here to denote both the political orientation of a work and its orientation, as Benjamin puts it, 'with regard to the degrees of technical development'. Benjamin does not address the problem of how the literary tendency of a work can be ascertained, which he considered a paradox in *The Concept of Art Criticism* (Section 2.3.1). The significance of his central claim, at any rate, lies in the implication that only the artistically progressive work can be politically progressive. Two things are important here. First, Benjamin tries to overcome the aestheticist drawbacks which he sensed in various otherwise appreciated varieties of modernist art and poetics, for instance in Valéry's constructivism or the surrealists' failure to proceed from ecstatic revolt to political relevance. He therefore argues that, in principle, the artistically progressive *can* be the politically progressive. Second, he opposes literary realism and any conception of political art that focuses on the content or message of the works. We can express this intention by emphasising that *only* the artistically progressive work, according to Benjamin, can be politically progressive. This aspect of his argument is directed against the already well-advanced communist policy, in 1934, of binding artistic production to the tradition of realism.[42] In its implication that the communication of particular contents is not important, however, it also agrees with his postulates of linguistic and literary self-reference. This abstract agreement finds a concrete expression in the form of Benjamin's argument.

He introduces the concept of literary *Technik* as follows: '[...] before asking, what the position of the literary work is *with regard to* the relations of production in its epoch, I would like to ask, how is it positioned *within* them. This question aims directly at the function which the work possesses in the literary relations of production of an age. In other words, it aims directly at the literary *Technik* of the works.' Benjamin does not define 'literary *Technik*' but seems to use the expression as an equivalent of 'the position of the literary work *within* the relations of production'. Any identification of the technique of a work with its position in what for Marx makes the structure of a society, however, would *presuppose* what Benjamin claims to demonstrate, namely that literary techniques have a broader societal and political significance. In the first sentence, moreover, Benjamin refers to relations of production in general; in the second only to literary relations of production. This shift from the Marxist concern with the production of material goods to the critic's concern with the production of literature is not explained and can be interpreted as a tacit equation of the two concerns.[43] Again, Benjamin risks losing his case before arguing it.

Benjamin's introduction of 'literary *Technik*' does not amount to the premature collapse of his argument only if literary production ultimately displays the same characteristics as non-literary production. Neglecting possible limitations in the Kantian dichotomy between interestedness and freedom from purpose, we would have to assume either that art is ultimately purpose-oriented or that production outside art is essentially non-instrumental. Adopting the first view means to reject the notion of the 'autonomy' of art. Benjamin's position on this point is unclear, for he argues for the importance of *literary* quality while affirming the dissolution of literature.[44] Moderately interpreted, the second view denies that man's use of matter and technology to fulfil his needs is an essential feature of economic production. Benjamin explicitly advocated such a denial both before and after 1934.[45] Only if we add a non-instrumental understanding of technology to Benjamin's explicit argument in 'The Author', does the introduction of 'literary *Technik*' not prevent the intended demonstration of his central claim.

The non-instrumental notion of *Technik* on which Benjamin implicitly relies conforms to the inclusion of 'Technik' in his early 'theory' of language. The new *physis* of mankind organises itself in *Technik* just as, in 'On Language', the spiritual essence communicates itself in language. *Technik*, to modify McCole's helpful observation, is the medium in which the new emerging collective *darstellt* ('prepares'), forms or 'organises', itself.[46] To the same extent, however, that Benjamin presents *Technik* as both the agency and the object in this process, he also presents it in accordance with (L) as a medium in the 'purest sense': *Technik* organises itself in itself. In 'Response to Oscar A. H. Schmitz' (1927), the otherwise deeply hidden political tendency of art is said to become manifest, 'as if of its own accord', in each technological revolution (GS II.2, 752; SW II, 17). Given that the political tendency is directed at, and thus defined by, that free interaction between man and nature which is the true *telos* of *Technik*, this conception implies that *Technik*, in technological revolutions, both reveals and organises itself. According to 'The Work of Art', it is the technological apparatus of film production that in film is communicated to the audience in view of its incorporation into the new collective techno-body (GS I.2, 487–8). *Technik* – or an invisible hand in *Technik* – seems to be in control of this potentially beneficial process, for the same technicalisation which in modern society drives the masses crazy is said to have created, for example in the films of Walt Disney, a vaccine against the craze (GS I.2, 462). And traditional art forms like painting and literature have a share in the ruse of *Technik* by calling, in their avant-gardist manifestations, for the technological revolution of film (GS I.2, 463–4). This implicit alignment of *Technik* with (L) invited the criticism that Benjamin advocated 'an ultra-modernist technologism' or conceived *Technik* as a historical agency operating independently of human action.[47] On the whole, however, the replacement of 'mankind' with 'Technik' remains as suspended in his later conception of modern culture, as did the equation of spiritual with linguistic essence, or language, in 'On Language'.

In 'The Author', the distinction between the position of the literary work '*with regard to*' and '*within*' the (literary) relations of production can recall that between communication *through* and communication *in* in 'On Language'. The repeated use of the adverb 'unmittelbar' ('directly', 'immediately') before the first occurrence of 'Technik' points in the same direction (GS II.2, 686; SW II, 770). It is Benjamin's subsequent assessment of that concept, however, which clearly shows that his notion of 'materialist' analysis conforms to that terminological mysticism which accompanies the postulates of self-reference also in his early work. 'With the concept of *Technik*, I have named that concept which renders literary production accessible to an immediate sociological and thus materialist analysis. [...] This concept, moreover, contains the instruction for the correct determination of the relationship between tendency and quality about which we inquired at the outset.' Benjamin may have 'named' the concept *Technik* but did not – and will not in 'The Author' – place it in the context of a (discursively developed) theory. For any discursive understanding of 'theory', no single term as such can contain the instruction to specify the relation between two concepts, or establish the functional dependence of political tendency upon literary quality. Only the contextualisation of a term might endow it with the far-reaching methodological powers that Benjamin attributes to 'Technik'. If in contrast one believes in mystic terminology and its attempt to combine a minimum of discursive thinking with a maximum of systematic importance (Section 2.3.2), the naming of the concept *Technik* may indeed seem to provide the key for the demonstration of Benjamin's central claim, or rather, it may appear to be a medium in which the truth of that claim is revealed. The double meaning of the German word *Technik* may then be the royal road to 'materialism', just as the ambiguity of *logos*, according to and in 'On Language', is indispensable for 'linguistic theory'. Benjamin's wording confirms his continuing reliance on a Schlegelian, non-discursive notion of analysis by the characterisation of the envisaged analysis *not* as immediate*ly* sociological, but as *immediate* and sociological, and thus 'materialist'.[48]

It is characteristic of linguistic mysticism that it replaces substantive discussion with 'terminological' decisions. Benjamin proposes that political correctness implies literary quality but fails to provide a substantial notion of literary quality that could support this claim. Instead, he refers us back to a passage according to which the literary tendency, the progress or regress, that is, of artistic *Technik*, 'makes' ('macht') the literary quality of a work (GS II.2, 685; SW II, 769). Oscillating between the senses of '*is*', 'represents' and 'generates', this use of 'makes' is reminiscent of his early endorsement of Schlegel's 'chemical' notion of criticism as 'preparation' (Section 2.3.2). Because we are not told why a work should be considered good if it is technically advanced, however, Benjamin's position threatens to become undistinguishable from an aestheticist affirmation of artistic innovation for the sake of artistic innovation.

The Scylla of aestheticism finds its complement in the Charybdis of an inartistic politicisation of art. At various points of his discussion, Benjamin comes very close to defining literary quality in political terms. He introduces Brecht's notion of a functional transformation ('Umfunktionierung') of literary institutions, on which he strongly relies, on political grounds and criticises the authors grouped under the heading of 'New Sobriety'[49] for abstaining from 'alienating, through improvements, the apparatus of production from the ruling class in support of socialism' (GS II.2, 691–2; SW II 773–4). In a related context, he provides with the degree of access to literary activity a criterion for the assessment of literary work that at least seems to be an exclusively political one (GS II.2, 696; SW II, 777). We can sympathise with a political understanding of art in general and with Benjamin's attempt to overcome his earlier cultural elitism in particular, but in the context of his argument, the political character of his assessment is precisely what is problematic. A political definition of literary quality begs the question whether or not the correct political tendency of a work implies its literary quality. It reduces Benjamin's central claim to a tautology.

The following comment on *Umfunktionierung* is perhaps even more problematic: '[...] the task of transforming the concert is impossible without the participation of the word. This participation alone, as Eisler puts it, can accomplish the transformation of a concert into a political meeting. And Brecht and Eisler have demonstrated the fact that such a transformation indeed represents a peak of musical and literary technique with their didactic play *The Measures Taken*' (GS II.2, 694; SW II, 776). If Benjamin claims the politicisation of artistic institutions to 'represent' ('darstellen') a maximum of artistic *Technik* because literary quality '*is*' or 'consists in' *Umfunktionierung*, he conceives literary quality in political terms and renders his main claim tautological. If, in contrast, we correctly interpret 'represents' in the weaker sense of 'amounts to' or 'necessarily implies', we are left with a repetition of Benjamin's main claim, which now, however, is said to be demonstrated not by him but by *The Measures Taken*. Let us concede that *The Measures Taken* is both politically correct and an artistic achievement of the highest order. The play does not therefore demonstrate that it possesses a maximum of literary quality which, moreover, is contained in its political tendency. In order to do that, it would have to provide both a criterion for the assessment of its literary achievement and a literary theory according to which this achievement partakes of the larger cultural and political transformation which it itself envisages. In crediting the play with the demonstration of his claim, that is to say, Benjamin conceives it as a piece of self-reflective art along the lines of his thesis. In 'The Author', *The Measures Taken* occupies that place which Goethe's *Wilhelm Meister*, as interpreted by Schlegel, occupies in *The Concept of Art Criticism* (Section 2.3.1).

Benjamin refers to Sergei Tretiakov's notion of the 'operating writer' to evoke the scenario of an immense process in which literary forms are being recast ('gewaltiger Umschmelzungsprozeß literarischer Formen').[50] This process, he argues, requires 'us' to recognise the changing importance of particular literary genres and different disciplines in the history of world culture. Commentary and translation, for instance, were not marginal in Chinese and Arab literature. In contemporary Europe, science and literature, criticism and 'production', education ('Bildung') and politics coexist unconnected and chaotically in a literary confusion, the setting of which is the newspaper.[51] The newspaper engages in an 'unscrupulous humiliation of the word' in which, however, the salvation of the word is being prepared. Benjamin's formula of the restitution of literature in the press of Soviet Russia comprises the replacement of depth ('Tiefe') with breadth ('Breite'), the abolition of the distinction between author and public and the literarisation of social life ('Literarisierung der Lebensverhältnisse'). Literary authority ('Befugnis') no longer rests on specialisation but is based on polytechnical education and turns into common property.

To a large extent, this image of an 'immense' cultural transformation is a romantic one. The notion that society ('Gesellschaft') adopts the character of literature or turns 'poetisch' figures prominently in the ideal of universal *Poesie* which Schlegel construes in *Athenäum* fragment 116. Like Benjamin's, Schlegel's literarisation of society coincides with *Poesie* or literature turning 'social' ('gesellig'), the transforming fusion of different genres, art forms and scientific disciplines and their formation ('Bildung') in and of a new order (KA II, 182–3). Benjamin's insistence on a connection between criticism and 'production' recalls Schlegel's postulate of the unification of criticism and *Poesie*, including its terminological presentation for instance in *Athenäum* fragment 238 (KA II, 204). The statement that *Umfunktionierung* requires the writer to work on the *means* of production, accordingly, both aligns Benjamin's image of cultural transformation with Schlegel's notion of 'transcendental' literature and removes it from any 'political materialism' for which revolution requires a change in the *relations* of production (GS II.2, 696; SW II, 777). Finally, the dissolution of fixed groups of producers and recipients is anticipated hermeneutically in the Romantic notion that the reader must be an 'extended' author.[52]

Benjamin does not argue that the Romantic ideal of universal *Poesie* is (being) realised in contemporary Soviet culture. In accordance with the normative use of 'culture' or 'philosophy' in his early writing, he proposes a romantic 'formula' for a restitution of literature that might be realised in the Soviet Union.[53] This formula, quite legitimately, is in part designed to accommodate his own economic and spiritual interests as a writer: commentary and translation, the forms of writing which he practised and intensively spiritualised, may rise to new prominence; even the newspaper for which the exiled, impoverished and 'proletariatised' author is forced to produce may become the place of a paradoxical – 'dialectical' – redemption of 'the word', springing, like the new religion desired in 1912, from a deep humiliation or 'enslavement'. Violently expelled from his home ground in German culture, Benjamin in 1934 no longer situates the desired revival in a German, hopefully cosmopolitan, 'youth', or in a reversal of 'allegory' decipherable from the German *Trauerspiel* of the seventeenth century but in a future 'Soviet Union' and imaginary 'proletariat'. The desired restitution, accordingly, is predominantly presented in terms not of religious awakening or spiritual revival but of 'technical' reorganisation. Yet the secular, polytechnical scenario to which Benjamin refers does not provide a 'literarisation' of social life in the required sense. A broader use of writing for economic or political purposes can increase literacy, technological progress and economic output but will not restitute an intrinsic and violated dignity to 'the word'. Such restitution requires either a 'literarisation' which leaves an auratic, ultimately religious, meaning of 'literature' intact or one in which the historical destruction of auratic literature somehow coincides with the messianic advent of the restituted 'word'. The 'linguistic' notion of *Technik* in 'On Language', which envisages the dissolution of depth ('Tiefe') in the openness of revelation, is compatible with the idea of such a coincidence (Section 1.3.2). It is in his 'Moscow Diary', however, that Benjamin expresses the hope that with 'Breite', or the very broadening of 'instrumental' communication, that point of destruction will be reached at which revelation may occur: 'Here it was necessary to evoke [...] the "destruction of language" as a tendency of present-day Russian literature. For the reckless development of what is communicative in language doubtless leads to the destruction of language. And in a different way the elevation into the absolute of its expressive character ends there, too, namely in mystic silence.'[54]

3.4.2 Epic illumination

Although Benjamin clearly did not wish the immediate audience of 'The Author' to recognise any religious meaning in his address, his reference to the salvation of 'the word' ought not to be dismissed as insignificant, merely nostalgic or an awkward expression of areligious concerns. For as I shall now try to show, Benjamin construes his main model of literary 'production', Brecht's Epic Theatre, at least in part as a forum for religious exercise. Brecht himself clear-sightedly noted the mysticism in Benjamin's concern with the technological destruction of aura and, of course, rejected it.[55]

In his radio talk on Brecht of 1930, Benjamin argues that Brecht's Keuner stories teach that 'new attitude' the teaching of which, according to 'The Author', makes Epic Theatre an outstanding model for *Umfunktionierung*. For Benjamin, Herr Keuner is 'the [...] one who belongs to all, that is, the leader. He is [...] by no means a rhetorician, a demagogue, a show-off or a muscleman. For Herr Keuner is the thinker.'[56] In contrast to Valérie's Monsieur Teste, Herr Keuner has a goal: 'This goal is the new state which rests on a philosophical and literary foundation as deep as that of the Confucian.' Benjamin here acknowledges Brecht's political intentions in the process of characterising them as philosophical and literary, and, indeed, as congenial to Asian religion. This spiritualisation and 'literarisation' of politics is also present in the use that he makes of Brecht's notion that gestures are 'quotable'. What is quotable about the attitudes of Brecht's characters, he maintains, 'is not only the attitude but equally the words that accompany it. These words, too, need to be practised, that is to say, first memorised, then understood. They have first their pedagogical effect, then their political, and their poetic last of all.' The political effect of the 'didactic' programme is said to precede, and seems to be given priority over, the poetic one, but in the pedagogical transmission of the required attitude words benefit from the aura of the poetic: they are treated as valuable and worth memorising independently of their (as yet inaccessible) meaning. However, in learning 'by heart' what still is incomprehensible, the pupil must also have faith in the truth of the teacher's words. The training programme for political 'organisation' envisaged here conforms to the requirements of religious initiation and exercise. Benjamin's 'thinker' Keuner teaches and 'leads' those who are to create or inhabit the new state – or '*in*' whose collective 'body' it is 'organised' – from a hierarchical position which rests on the aura of 'the word'. This suggests that at least in 1930 Benjamin's opposition to the post-expressionist 'Activism' of Kurt Hiller, whom he strongly attacks in 'The Author' for advocating a 'logocracy' of intellectuals, rests on the assessment that the *logos* the rule of which Hiller demands is not holy enough.[57]

In 1934, Benjamin no longer uses 'the leader', 'the thinker' or references to Confucius. He comments on the 'tests' and 'assessments' to which Epic Theatre subjects modern man as follows:

1. What emerges is this: the events are alterable not at their climaxes, not by virtue and resolution, but solely in their strictly habitual course, by reason and exercise.

2. The construction, from the smallest components of behaviour, of what in Aristotelian dramaturgy is called 'action' is the purpose and meaning of Epic Theatre. [...]
3. [The epic dramatist] intends less to fill the audience with feelings, even of revolt, but to alienate it lastingly, through thinking, from the circumstances in which it lives. [GS II.2, 698–9; SW II, 779]

(1) formulates an insight concerning the source of social change that may be gained from Brecht's work: moral effort will not do the job. The positive suggestion that the application of 'reason' and 'exercise' to human habits will – and not a direct change of external conditions and social structures – is perhaps less Brechtian.[58] At any rate, (1) provides Benjamin's view of what is required to alter the course of events. (2) defines the central intention of Epic Theatre, and (3) describes the effects intended with regard to the audience, to the people, that is, who are to adopt the new 'attitude'.

(2) is difficult to understand. There are at least three things that Aristotle considers 'action'. First, human practice (*praxis*) as such, which he takes to be represented on stage; second, the means of such representation, which is accomplished through the acting (*drōntōn*) on stage as opposed to epic narration; and third, the plot (*mythos*) of the play.[59] If Benjamin uses 'action' in one of the two latter senses, he conceives Epic Theatre as an exercise in self-reflective literature. Accordingly, Epic Theatre would intend the analytical and detailed 'construction', and presumably also the critical re-formation, of the plot or acting of traditional drama. This understanding of Epic Theatre may be correct, and it may also be correct to attribute it to Benjamin. However, what is needed in 'The Author' is an argument in support of the political relevance of such a theatrical exercise in theatre criticism. Since no specific argument to this effect is presented, I assume that Benjamin at least in part uses 'action' in the sense of 'human practice', with that difference from the tradition, of course, that human practice is considered changeable.[60]

(2), then, implies the claim that the purpose and meaning of Epic Theatre lies in the construction of human practice from the smallest components of human behaviour. What does 'construction' mean here? Benjamin presents Epic Theatre as a 'laboratory' where 'the components of reality' are treated ('behandelt') in the sense of an experimental grouping ('im Sinne einer Versuchsanordnung') (GS II.2, 698; SW II, 778). If this use of scientific vocabulary is metaphorical,

Benjamin's notion of construction is unhelpful because he does not indicate in what way a metaphorical recourse to scientific creativity in the context of social change differs from mere moral encouragement. The unspecific hope that things can be changed, which the metaphor 'laboratory' conveys, belongs to those moral virtues the relevance of which is ruled out by (1). If, however, Benjamin's scientific vocabulary has to be interpreted literally, we are not dealing with an application of scientific attitudes to art and politics, but with a magical understanding of both. His terminology then suggests that in Epic Theatre reality and human practice are analysed and construed as substances are in a laboratory. 'Construction' in 'construction of human practice' would refer to a production, or 'preparation', of practice. In 'Epic Theatre' we would find that *Darstellung* which Benjamin found in Romantic 'criticism' (Section 2.3.2), except that now bodily representations of human action have to be included within those linguistic phenomena which are assumed to function as (living) matter. In scientific experiments, of course, the 'components of reality' that are used – the samples of the substances which are being analysed or manipulated – must be at hand. Images or representations of the components, or thought experiments, will not do. If Epic Theatre literally 'constructs' human behaviour as a chemist prepares a new substance, the components of human behaviour, and not merely representations of these components, have to be present on stage. The idea of a theatrical production of human practice, that is to say, rests on the 'magical' equation of the represented object with the representing medium (Section 1.2.2) and implies that the practice to be created and, by extension, the sphere of social action as a whole, are magical in kind. Let us call this interpretation of Benjamin's engagement with Brecht the 'magic thesis'.

The magic thesis can draw on considerable textual support from outside 'The Author'. In 'Madame Ariane: Second Courtyard on the Left', a text from his 'constructivist' *One-Way Street*, Benjamin considers a bodily enactment by which the meaning of signs is changed 'the true practice' (GS IV.1, 142; SW I, 483). According to 'Programme of a Proletarian Children's Theatre' (1928–29), it is the task of the director of the envisaged workshops to redeem the signals which emerge from infantile phantasies of omnipotence by rendering them effective in the application to matter and artistic materials ('zur Exekutive an den Stoffen zu bringen') (GS II.2, 766–7; SW II, 204–5). And in the surrealism essay, Benjamin finds the required political organisation of 'pessimism' in the discovery of 'the one hundred per cent image space' in the 'space of political action', in the discovery, that is, that the political sphere contains, or even '*is*', that of actions which are identical with the images which they produce (GS II.1, 308–9; SW II, 216–17). Political and human practice in general, it would appear, is a magical operation for which one can train in the theatre.

However, there are strong reasons for not embracing the magic thesis. First, the thesis implies that, when in the famous finale of 'The Work of Art' Benjamin opposes the 'politicisation of aesthetics' to the fascist 'aestheticisation of politics' (GS I.2, 469, 508), such politicisation would actually come down to a body-centred magical practice comparable to that enacted in some Nazi rituals. We ought not to affirm this implication without necessity. Second, I arrived at the magic thesis via the exclusion of a metaphorical use, in 'The Author', of scientific vocabulary. I excluded such a use because it would render Benjamin's discussion vague and his point hardly distinguishable from the explicitly rejected position of 'bourgeois' humanism. Benjamin's discussion, however, may indeed be vague. Also, a different metaphorical interpretation of his scientific vocabulary with less negative implications may be possible. Third, the magic thesis conflicts with Benjamin's explicit endorsement of Brecht's rejection of dramatic illusion (GS II.2, 698; SW II, 778). And finally, the thesis at least appears to be contradicted by Benjamin's recourse to 'reason' and 'thinking' in (1) and (3).

Benjamin names 'reason and exercise' but does not indicate how the two can alter the strictly habitual course of events. In particular, he does not discuss the role of reason in human action. Neither does he set up, as his scientific vocabulary might lead one to expect, a *behaviourist* scenario, in which man is subjected to an experimental training programme designed by the 'reason' of scientists or scientific dramatists. Apart from rather abstract uses of 'organisation' and its cognates,[61] and apart from the indirect evocation via Aristotelian dramaturgy, human action, habitual or not, is strangely absent from his picture of politicised art. The passage under discussion – not dissimilar in this regard to 'The Religious Position' (Section 1.3.3) – resonates with the traditional opposition of action and contemplation, and Epic Theatre comes to sit on the latter side of the fence: Aristotelian dramaturgy is concerned with 'action', Epic Theatre with 'thinking'. As we have seen, Benjamin's 1930 interpretation of Brecht can be described as religious and in particular suggests the relevance of a religious understanding of 'exercise'. In *The Concept of Art Criticism*, 'thinking' is construed as a medium in the form of which the absolute is present (and absent) (Sections 2.2.1–3). And Benjamin repeatedly characterises the modality of revelation with recourse to the verb 'vernehmen' ('to hear'), which is etymologically related to 'Vernunft' ('reason').[62] In the remainder of the present section, I shall discuss those features of Benjamin's 1934 construction of Epic Theatre which permit us to dismiss the magic thesis in favour of a religious interpretation, which can acknowledge the partial rationalisation of the magical elements in Benjamin's engagement with Brecht.

According to (3), the spectators of Epic Theatre are not alienated from their reality, at least not 'lastingly' enough, but ought to be. Their lasting alienation is achieved through the 'thinking' to which the epic dramatist exposes them. Benjamin illustrates this thinking with a so-called example ('Beispiel'), which he also presents as a comment on Brecht's techniques of montage and *gestus*. The relevance of the example to 'epic' thinking is underlined by the common element 'fremd' ('strange') in 'Fremder' ('stranger') and 'entfremden' ('estrange', 'alienate'), and the fact that the stranger, to whom Benjamin compares the epic dramatist, interrupts the process and thus 'alienates' the audience from the represented action. If we disregard for the time being a remark on the importance of laughter,[63] Benjamin's example contains the only explanation of 'thinking', and the related 'reason', in 'The Author'.

> Stellen Sie sich eine Familienszene vor: die Frau ist gerade im Begriffe, eine Bronze zu ergreifen, um sie nach der Tochter zu schleudern; der Vater im Begriff, das Fenster zu öffnen und um Hilfe zu rufen. In diesem Augenblick tritt ein Fremder ein. Der Vorgang ist unterbrochen; was an seiner Stelle zum Vorschein kommt, das ist der Zustand, auf welchen nun der Blick des Fremden stößt: verstörte Mienen, offenes Fenster, verwüstetes Mobiliar. Es gibt aber einen Blick, vor dem auch die gewohnteren Szenen des heutigen Daseins sich nicht viel anders ausnehmen. Das ist der Blick des epischen Dramatikers. [GS II.2, 698; SW II, 778–9]

> Imagine a family scene: the wife is about to seize a bronze sculpture and throw it at her daughter; the father is about to open the window and to call for help. At this moment, a stranger enters. The process is interrupted. What comes to light in its place is the state on which the stranger's gaze now falls: stricken faces, open window, destroyed furniture. There is a gaze, however, before which the more usual

scenes of present-day existence do not look very different: that of the epic dramatist.

As presented here, alienating 'thinking' is a particular kind of perception. Benjamin's 'epic dramatist' *sees* things in a certain way. His perception, which he tries to communicate to the audience by the use of montage, both follows and rests on an interruption in the course of events. The arrest of *Vorgang* ('process') brings to light a *Zustand* ('state') in which the truth underlying the *Vorgang* may be revealed. The envisaged 'thinking' or 'attitude', that is to say, imitates in its very form Benjamin's notion of messianism and corresponds to that 'arrest' or *Stillstellung* of thoughts that in 'On the Concept of History' will be attributed to the 'historical materialist'.[64] In the devastation of the bourgeois *interieur*, moreover, the *Zustand* on which the stranger's gaze is said to fall displays that deformation ('Verbildung') in the present from which, according to 'The Life of Students', the future ('das Künftige') has to be liberated in an act of cognition ('erkennend') (GS II.1, 75; SW I, 38). No sign of help is present, only a gap in the boundary of the *interieur*, an open window, or so we are told. (The window may actually be closed since the father was only about to open it when the process came to a standstill.) The colon between the word 'state' and its complement 'on which the stranger's gaze now falls' on the one hand, and the list of physically perceivable objects on the other, suggests that the *Zustand*, too, can be physically perceived. The same effect is achieved by the choice of 'before which' ('vor dem') instead of 'for which' ('für den') in the last sentence. However, there is no ascertainable physical 'gaze' by which a complex state of destruction-liberation could be perceived. What we know to exist, in contrast, is the mystic device which Benjamin uses here: the evocation of a literal understanding of those metaphors of perception with which the mystic expresses the intuitions, anxieties and desires which he experiences – at times 'bodily', we may assume – *in the context of* external perception. Such evocation (or its theatrical 'realisation'), of course, suggests the objectivity of mystic experiences and may even trigger similar ones in a reader (or audience). The 'thinking', then, by which Benjamin's epic dramatist alienates the audience from ordinary, everyday modes of perception is neither thinking nor external perception, but a communication of mystic experience which conforms both structurally and symbolically to his idea of messianism. If we take his example seriously, we have to conclude that the successful propagation of 'epic' alienation outside the institutions of bourgeois culture would lead to an increase of religious 'practice' in a society, and that this might well be the only *direct* alteration of the 'habitual course of events' which it can achieve.

The outlined interpretation, however, neglects the grotesque and comical features of Benjamin's example. The arrested throwing of a bronze sculpture in what after all may be a banal family row may well provoke laughter and conflict with the evoked depth of mystic vision. In its comparisons – between the epic dramatist and the stranger, between a family scene and 'present-day existence' – Benjamin's example resembles a parable. The tension between the seriousness of visionary depth and the comical, in turn, points to his understanding of Kafka's parables as formulated in the 1934 Svendborg notes.[65] In 'The Author', that is to say, Benjamin uses a literary (and religious) mode of expression to illustrate the effect of Epic Theatre on the audience. He conceives what would be the central link between theatrical and political practice in terms of a literary practice which, moreover, is given a clearly religious interpretation in his work on Kafka during the same period. With regard to 'The Author', Benjamin's Kafka essay represents a helpful source. In the former text, the significance of the example of the stranger for the notion of *gestus* remains obscure. Nothing in Benjamin's *tableau* points to class antagonism, which for Brecht is the underlying social structure in many gestures that makes *gestus* a valuable *technique*.[66] In the two essays 'What is Epic Theatre?' of 1931 and 1939 respectively, Benjamin tends to present *gestus* as an exclusively formal concept, closely linked to 'montage' and 'quotation' (GS II.2, 521, 535–6). What is required in 'The Author', however, is a substantial notion of *gestus* that relates the theatrical technique to human practice in general. In 'Franz Kafka', we find such a substantial notion of *gestus*. Only, it concerns not political but religious practice.

Kafka's work, Benjamin argues, is theatrical, it presents a 'code of gestures' which Kafka probes for a hidden meaning in ever-changing contexts and, in the expression from 'The Author', 'experimental groupings' (GS II.2, 418–20, SW II, 801–2). In 'Franz Kafka', the use of the scientific phrase is clearly metaphorical and leads us not to 'bourgeois' humanism but into that realm which already the early Benjamin opposed to secular humanism, religion. With recourse to the motif of the door-bell in Kafka's 'A Fratricide', Benjamin characterises Kafka's literary technique as follows: 'As this bell, too loud for a door-bell, rings out towards heaven, the gestures of Kafka's characters are too forceful for the usual environment and break out into wider areas.'[67] The excess which gestures can display in the usual, everyday environment, then, may have a potentially religious meaning. The arrested throwing of a bronze sculpture in 'The Author' is excessive enough; the usual plate, after all, does not do for Benjamin's family row. If, furthermore, the religious potential of excessive gestures can be compared to that of a door-bell which is too loud for a door-bell, it can also be found in an open window, a small gap in the confines of the oppressive *interieur*, which may provide room to 'break out into wider areas'. (Especially so, if the window may not be open as yet.) In 'The Author', Benjamin not only presents the 'epic' technique of montage on the basis of the contrast between 'Vorgang' and 'Zustand' which is central to his messianism, but also chooses an example to characterise Brecht's technique of *gestus* which accords with Benjamin's own negative-theological interpretation of Kafka. What Benjamin finds in the symbolic experiments that are Kafka's gestures are so many prayers for revelation; post-traditional prayers, of course, since they take place in literature and not in the synagogue. Perhaps the laughter with which Kafka's texts resonate – and which is so absent from most of Benjamin's – is a post-Judaic one. It is certainly (also) an appropriate outlet for the scepticism of praying sceptics.

Benjamin's interpretation of Kafka's 'gestures' agrees both with Kafka's notion of writing as a form of prayer[68] and with the performative mysticism of Benjaminian 'theory'. Benjamin's engagement with Brecht's ostensibly secular art can be described as a probing for religious meaning where revelation is unlikely to occur. Against the background of a Brechtian understanding of Brecht, such an 'experiment' appears grotesquely misplaced. Whenever (political) action is needed, moreover, it will appear an exercise fraught with danger in the eyes of anyone for whom prayer is not, as Benjamin held in 1919, a different kind and higher 'degree' of action. In Benjamin's work, however, his 'experiment' with Brecht represents merely a more complex, 'theoretical', counterpart of attempting what he chose to call 'profane' illumination: the religious illumination '*in*', and of, the secular.

3.4.3 Contemplation, film, politics

In 'The Work of Art', Benjamin extends his experiment with Epic Theatre to the technology, techniques and reception of film.[69] Although he dismisses other authors who relate film to the sacred or supernatural as reactionary (GS I.2, 486–7), he alludes to the theological *topos* of a restituted and 'speaking' nature: 'Thus it comes within one's grasp that the nature which speaks to the camera is different from that which speaks to the naked eye' (GS I.2, 500). Used not in the idiomatic sense of 'to turn violent' but in the literal sense of 'to come within one's grasp', the expression 'handgreiflich werden' links the theological *topos* to the centre of Benjamin's theory of film and twentieth-century modernity. For the tensions of 'our time', he argues, lead to the dominance of the tactile even within the realm of the optical, and this is the case, above all, in the shock effect of films (GS I.2, 505). By presenting the *topos* of a new and speaking nature as something the truth of which becomes *handgreiflich,* Benjamin stresses that both the visual metaphors of mysticism and the optical achievements of the cinema are insufficient to communicate the envisaged restitution. In his conception of historical crises, accordingly, he rejects 'contemplation' in favour of 'tactile reception': *'The tasks which the human apparatus of perception faces at the turning points of history can in no way be solved by merely optical means, that is, by contemplation. They are mastered slowly by a formation of habits modelled on tactile reception.'*

The contemplation that is rejected here, however, is not the opposite of an active life, for it is the human apparatus of perception, and not human action, which is said to face historical tasks. The reduction of contemplation to the use of one's eyes or optical equipment, moreover, is misleading at least to the same extent that Benjamin's interpretation of revelation as something that occurs in a *Vernehmen,* or intensive 'taking',[70] is appropriate. Leading an active life, on the other hand, involves more than getting used to things or employing the sense of touch. In what follows I shall argue that Benjamin's 'theory' of film is designed to reaffirm a non-reductionist notion of contemplation, to reaffirm, that is, an originally religious experience which affects the *whole* body of a person who abstains from active involvement in social, secular practice. To put it differently: in central aspects of its so-called materialism, 'The Work of Art' agrees with that note for the Arcades Work which states that contemplation has to be put on trial but will 'defend and assert itself brilliantly' (GS V.2, 1036).

The home of 'tactile reception', according to Benjamin, is not film but architecture (GS I.2, 504–5). Buildings, he states in partial accordance with the classical view of architecture as a borderline case of art,[71] are received 'by use and by perception, or rather, by touch and sight' (GS I.2, 465, 504). Thus 'tactile reception' seems to designate not the reception of a work of art as art, but the use of an object, which may also be a work of art. Yet 'tactile reception' is a peculiar phrase[72] and Benjamin's explanation of it conflicts with what is usually considered the use of a building. We do not use (or get used to) a building by touching it, at least not exclusively; 'by touch' is *not* a better expression for 'by use'. Nor do we seem to 'receive' an object that we use. We do not 'receive' it as art, as Benjamin implies in accordance with Kant, but neither do we 'receive' it in any other obvious way. 'Tactile reception' is peculiar since it seems to focus on the physical contact between the user and the object and disregards the social practice in which we use, and get used to, buildings. In Benjamin's model for that habit formation which solves historical tasks, however, such apparently physical contact is conceived in terms of an incorporation, into the subject, of the object. This incorporation, he claims, forms the opposite of contemplation: 'Distraction and contemplation form an opposition which permits the following formulation: the person who concentrates before a work of art is absorbed by it; he enters into this work as legend tells of the Chinese painter who looked at his completed work. In contrast, the distracted mass on its part absorbs the work of art and, most obviously, the buildings' (GS I.2, 504).

The more literally we interpret the 'absorb' ('in sich versenken') of the last sentence, the more the work of art which the mass is said to absorb comes to fulfil a function similar to that attributed to the shared food in religious rituals. The members of the mass would not only be distracted ('zerstreut') but also dispersed ('zerstreut'), and the 'absorption' of the work would unite them in that 'bodily collective' which according to Benjamin's notes on surrealism must be the basis of 'materialism'.[73] Understood as a metaphor for reception, in contrast, the 'absorption' of the work attributed to the mass is still a form of *Versenkung* ('absorption', 'contemplation'). It may be a collective and less voluntary *Versenkung* than that of the bourgeois concentrating before a work of art, but Benjamin's indications are insufficient to consider it different in kind. Since any reception of a work is a reciprocal process between the work and those receiving it, the contrast between the absorption of the work by the mass and the absorption of the concentrating beholder by the work is unconvincing. In his 'Moscow Diary', Benjamin indirectly confirms the view that 'tactile reception' is not different in kind from contemplation when describing his empathy in a Cézanne painting in precisely those terms that in 'The Work of Art' are reserved for the 'tactile' opposite of contemplation (GS VI, 325). 'Tactile reception', then, is a construct from which any social interaction is absent which does not either resemble religious practice or is a variety of aesthetic contemplation. In presenting 'tactile reception' as a progressive social tendency, Benjamin advocates a non-institutional respiritualisation of modern society. Accordingly, he praises the decrease of the power of the church in modern times as an achievement, but considers it a sign of decline that 'those forces which the individual puts to work in his relations with God' have been withdrawn from public life (GS I.2, 502, note 27).

Although film is not particularly related to the sense of touch, Benjamin conceives the cinema on the model of 'tactile reception'.[74] He claims that as the mass absorbs the building, the spectator absorbs, or identifies with, the technical 'apparatus' of film production. Adopting the 'attitude' of the apparatus, the spectator tests the artistic achievement of the actor, namely his adaptation to the requirements of (film) technology. The rapid interruption in the sequence of film images, moreover, produces a *physical* shock in the spectator and thus conforms to the dominance of the tactile in contemporary perception. The shock effects of film solicit or require a heightened presence of mind. Film thus becomes an instrument of exercise: the spectator practices appropriate responses to the increased threats to his life that he faces in the metropolis and the political situation of 1936. Finally, film also becomes an instrument of control. Surreptitiously ('unter der Hand'), through film art tests the extent to which the spectator has formed habits appropriate to solving the new tasks of perception.

This 'theory' of film reception has frequently been criticised as psychologically implausible and empirically inaccurate.[75] However, the overall picture that emerges from Benjamin's claims is of a kind that makes such criticism, correct, as it may be, nearly misleading. The 'identification' with an apparatus and the exposure to shock, adaptation exercises and controls lead, as Benjamin knows (GS I.2, 462), to trauma and alienation, or to boredom. No presence of mind springs from the exposure to shock. We attribute 'presence of mind' to a person who 'does the right thing' to ward off danger, while others may be paralysed by shock. Sometimes, the very person who acts present-mindedly is seized by shock after the event. If present-mindedness can be trained at all, any such training would seem to require that the trainee learns to focus her attention on practical needs in a situation of danger, and not that she is exposed to shock in a state of distraction.[76] The assumption that the rapid interruption in certain types of montage causes physical shocks, which can be compared to those experienced in mortal danger, is implausible. Benjamin presents the cinema as an experimental setting in which the spectator both tests artistic achievement and is tested for her level of shock management. Yet it is difficult to imagine any 'laboratory' where the two tasks could be carried out at the same time.

Benjamin is not concerned with empirically observable events in a laboratory or cinema. He proposes a normative conception of film that to a large extent is shaped by his specific understanding of the requirements which the respiritualisation of modern society must meet. Those features of his 'theory' that in an empirical interpretation are implausible if not absurd form a meaningful (religious) pattern if we relate them to the spiritual concerns of his 'anthropological materialism' and rhetoric of self-reference. I shall try to show this with regard to, first, his claims concerning the relation between interruption, shock and 'presence of mind', second, his positive evaluation of the spectator's identification with the apparatus, and last, his implication that the spectator is both the subject and the object of tests and controls.

1. If at all, the rapid interruption in the sequence of film images does not produce shock in the spectator independently of the latter's disposition. The spectator whom Benjamin presupposes in this context is trained in 'contemplation'. Benjamin quotes Georges Duhamel's complaint 'I can no longer think what I want to think. The moving images have taken the place of my thoughts' and comments as follows: 'The spectator's process of association is indeed interrupted by the rapid change of the images that he sees. On this interruption rests the shock effect of film, which, like any other shock effect, has to be parried by a heightened presence of mind' (GS I.2, 503). The rapid change of images leads to shock in someone who 'want[s] to think'. Duhamel's complaint registers the loss of autonomy that his thought suffers in the cinema. The distraction that he experiences does not equal the distraction that the masses are said to seek (GS I.2, 504). He is distracted ('abgelenkt') from something he wants to do, thinking for himself, while we, when seeking distraction ('Zerstreuung'), avoid something we do not want to do, which at times includes thinking for ourselves. Benjamin's undifferentiated use of 'Zerstreuung' to characterise *one* mode of modern reception is therefore misleading. For distraction to be experienced as *physical* shock, moreover, the 'thinking' that is interrupted by the change of images must be experienced with an extraordinary bodily intensity. It must resemble the mystic's highly charged 'contemplation' and differ from any relaxed variety of aesthetic reception.[77] At this point, we have left Duhamel, who sounds annoyed – and arrogant – but not literally shocked.

Benjamin's early 'Schemata on the Psycho-Physical Problem' tries to objectify the mystic's bodily experience of his thinking in 'anthropological' categories. In one dimension of the schemata, spirit is identical with the body; it is, emphatically, 'bodily spirit' ('leibhafte[r] Geist'). The spiritual nature of living beings is also seen as closely bound up with that *corpus* which belongs to God, and the resurrection of which coincides with the dissolution of bodily ('leiblich') life (GS VI, 78–81; SW I, 393–5). In the fragment 'On Horror I' (1920–22), Benjamin accordingly conceives *Geistesgegenwart* as that presence *of spirit* ('Gegenwart des Geistes') and 'holy absorption' in God which offers the best protection against horror.[78] 'On Horror II' (1920–22) links the absence of language ('Sprachlosigkeit') to horror (GS VI, 77). *Geistesgegenwart*, we may therefore say, here coincides with 'language' as conceived in 'On Language' (Sections 1.2.2–3). In later publications which discuss the most intimate relation between body and spirit under the heading of 'Geistesgegenwart', the connotation of 'presence of spirit' is still there. According to 'The Path to Success in Thirteen Theses' (1928), it is only the body that can achieve presence of (the) mind: 'Gegenwart des Geistes schafft allein der Leib' (GS IV.1, 352; SW II, 147). And in 'Madame Ariane', 'bodily presence of mind' is presented as the result of ascetic exercises – such as fasting, chastity and vigil – and credited with the realisation of the 'only desirable telepathic miracle', namely 'to turn the threatening future into the full *now*' (GS IV.1, 142; SW I, 483).

Benjamin's obvious interest in the bodily presence of mind shows that he either at times experiences thought processes with bodily intensity or considers such experience important in principle. This explains that he can construe the distraction from a line of thought as a physical shock. He thereby implicitly assumes the spectator to have adopted a mode of intensive 'thinking' which resembles that of the mystic's 'bodily' contemplation. Placed against this background, his insistence on a heightened presence of mind amounts to the view that the bodily intensity of internal thought experience, which explains the occurrence of shock in the first place, has to be increased even further. This view makes sense to the same extent that it is 'the full *now*' which is being desired, for any revelation of intensive infinity certainly benefits from the acceleration and intensification of bodily experienced thought processes. If 'the full *now*' is the aim, the shocked mystic who practices heightened presence of mind in the cinema '*does* the right thing' without having to lift a finger.

However, in 'Madame Ariane' Benjamin also relates 'presence of mind' to externally observable bodily responses: 'The ancients still knew the true practice, and Scipio, stumbling as he set foot on Carthaginian soil, cried out, spreading his arms wide in the fall, the password of victory: Teneo te, Terra Africana! He bodily binds what was about to become an omen of disaster and a sign of misfortune to the present second, making himself the factotum of his body.' By spreading out his arms to embrace African soil, Scipio urges the Romans to interpret his fall, which otherwise would have been received as a bad omen, as a sign of hope. Without time for reflection, he does the right thing to keep them convinced that their African adventure will succeed. Benjamin's interpretation of Scipio's mastery of the situation as an example of 'the true practice' is remarkable for what is absent from it. Does Scipio's turning of the threatening future into the full *now* imply that by his display of bodily presence of mind he ensures Roman victory? Is his presence of mind a 'telepathic miracle' because ultimately an agency which sends signs or the 'password of victory' determines the future course of events, or because that general will necessarily win who keeps his troupes convinced that they will? Or is it rather the case that any experience of the full *now* renders the question of who will win unimportant anyway? By not addressing these questions, Benjamin avoids any positive statement of those magic or religious assumptions that in some way or other are implied in his notion of 'true practice'. However, we can positively establish that considerations of two kinds are absent from this notion in 'Madame Ariane'. These are considerations of practical prudence and of political or moral value, for instance the questions which strategy should be adopted in the attack and whether imperialist conquests are a good idea in the first place.

Geistesgegenwart is nevertheless an eminently 'political' category for Benjamin. Just as with Scipio's fall, things happen too quickly in politics to analyse the situation first and decide on an appropriate course of action on the basis of rational assessment. When rational action in this sense could start, the situation will have changed. Therefore, the later Benjamin argues with repeated reference to Anne Robert Turgot, politics requires one 'to foresee, as it were, the present'.[79] In Benjamin's interpretation, Turgot's *bon mot* on the necessity of political intuition assumes considerable weight. The political 'present', he argues, is the intentional object of a prophecy. It is the intentional object of a prophecy, moreover, which in accordance with Schlegel's notion of the historian as a prophet turned backward proceeds from the past.[80] Weidmann has shown that along these lines, 'Geistesgegenwart' becomes an essential concept for both the performative dimension of the Arcades project and Benjamin's notion of the necessary political significance of any true historiography.[81]

According to Weidmann, Benjamin's prophetic historian even proves to be a politician. As the *intentional* object of a prophecy, however, Benjamin's 'present' is severed from reality, it is subjected to the phenomenological operation of *epochē*. The present intended in the true historian's prophecy may never be the political reality in which people live. The historian's prophecy, moreover, cannot have a direct impact on political action or even on political displays of *Geistesgegenwart*. If there is no time for rational analysis before action must be taken, there is of course no time either to write or read, say, a work on arcades and the 19th century. In 'Madame Ariane', Benjamin clearly rules out the possibility that the historical exegete of 'what happens in the present second' can become, by virtue of his 'prophecy', a present-minded agent (GS IV.1, 141; SW I, 483). It therefore seems that only two possibilities exist for the prophet's intended present to become the actual one. Either his intentions or thoughts cause corresponding bodily innervations in political agents, or the formation of the intentional object in the prophet's mind provokes a superior agency into action that has the power to realise the 'prophecy'. Benjamin's 'political' understanding of *Geistesgegenwart*, that is to say, comes as close to positive magical or religious assumptions as does his interpretation of the Scipio episode. In the case of 'political *Geistesgegenwart*', however, we can draw on strong evidence which suggests two things: first, that the notion of *Geistesgegenwart* which underlies Benjamin's 'theory' of film rests on a highly spiritual, and in this sense 'religious', understanding of 'politics'; and second, that it nevertheless does not necessarily imply the positive affirmation of traditional religious beliefs.

Benjamin did not first encounter the notion of reason's necessary belatedness in Turgot. Unger relates this notion to the requirement that spirit be as unmetaphorically 'present' and effective as is the body: 'For *Geist* to be effective it must be given the possibility t o b e p r e s e n t with the same directness, with the same unquestionable, unmetaphorical force and suddenness as is the effective body. Nothing else can guarantee that its perpetual belatedness will end.'[82] For Unger, the desirable overcoming of capitalism and, indeed, the replacement of the social 'pathology' of *all* history with an ethically satisfactory order requires the creation of the possibility of spirit affecting matter and physical events immediately. Spirit has to become a 'bodily' force. At the very outset of the Weimar Republic, Unger presupposes that any non-spiritual engagement in politics – no matter whether liberal or radical – is necessarily bound to result in the endless recurrence of historical catastrophe. The belatedness of rational intervention necessarily thwarts the realisation of rational objectives. In order to endow spirit with bodily force, however, a change in the psycho-physiological outfit of European man is required. Within modern European culture, there are only marginal traces of the required potency of spirit, in spiritualism, for instance, and perhaps also in the practice of hypnosis. Yet, however difficult, the introduction into European society of the spiritual and parapsychic practices of non-European or pre-modern cultures is *a priori* the only possibly successful option. Its realisation would abolish the modern separation between religion and the state and re-establish the only legitimate form of rule, namely the hierarchical one of spiritual leaders or priests. Prophetism would become the basis of political action and replace the belatedness of reason. Unger deals with the demand of a spiritual society as a practical 'problem'. He does not argue, that is to say, for the truth of magical and religious beliefs but considers the possible effectiveness of practices and social structures historically related to such beliefs as sufficient for his proposal to redefine 'politics' in spiritual terms.

In 1921, Benjamin could call Unger's *Politik und Metaphysik* the possibly 'most significant contemporary work on politics' because in his 'Programme of the Coming Philosophy' he himself had already advocated the revival of pre-modern or non-European types of spiritual practice (GB II, 127). The absence, in Unger's practical argument, of positive religious assumptions, furthermore, conformed to Benjamin's religious scepticism while the argument as a whole was suited to accommodate his desire for religion. With regard to the later Benjamin's understanding of 'politics', it is important to realise that in *Politik und Metaphysik* we find not Turgot's weak notion that the politician has to foresee, 'as it were', the present, but an unmetaphorical affirmation of 'political' prophetism, which moreover is combined with both the notion of reason's necessary belatedness and the requirement of a bodily presence of mind. We find, that is, a fully developed argument for the respiritualisation of modern society that combines three of the key concepts of the later Benjamin's concern with 'politics'. Placed against this background, Benjamin's insistence, in the 'Work of Art', on 'heightened presence of mind' aptly complements his disregard for any non-spiritual form of social interaction and any non-aestheticised 'use' of things. The shocked mystic who practices heightened *Geistesgegenwart* in the cinema 'does the right thing' even independently of the gratification that the experience of intensive infinity holds in stock. His training will not immediately make him fit to deal with the mortal dangers of modern traffic or Nazi persecution but he, for his part, embraces the only option available to end the eternal recurrence of historical catastrophes. Like the religious person leading a contemplative life, he pursues not any particular political objectives but the replacement of secular politics with spiritual practice. To the extent that the masses lead an active life, moreover, the required spiritualisation of society will not emerge from them. This is why, for Benjamin, the Duhamels among his contemporaries cannot be left alone: they *must* go to the cinema and be trained.

2. According to Unger, the psychophysical outfit of the human body has to change in order to permit the spiritual conduct of public affairs. With his positive evaluation of the spectator's identification with the apparatus of film production, Benjamin attempts to construe technology as instrumental in realising this condition of society's respiritualisation.[83] Given that technological development cannot simply be reversed, an attempt of this kind forms a necessary part of any modern project of spiritualisation. For, as Benjamin convincingly argues, modern technology threatens the human body with social insignificance.[84] It replaces by instruments, that is, what alone can achieve presence of (the) mind. Therefore the human body has to 'incorporate', or identify with, the apparatus of modern technology precisely if it is to regain its dignity as a possible medium of socially relevant revelation.

The apparatus of film production permits the spectator to see things that cannot be seen by the naked eye. In this, 'cinematic' perception resembles those parapsychic or spiritual 'experiences' that according to 'Programme of the Coming Philosophy' were lost in modernity but need to be regained and based on the notion of God.[85] Film technology, like the prophet or clairvoyant, breaks through the limits that the 'bourgeois' common sense notion of knowledge ascribes to human perception. To the extent that the spectator's eye really identifies with the camera lens – or his brain with the cutting and montage equipment – he will not only see different images but also believe that he is experiencing (with his whole body) what according to 'bourgeois' standards cannot be experienced. The collective 'reception' or incorporation in the cinema (GS I.2, 460, 497), moreover, ensures that his religiously fruitful experience is no longer enclosed within the confines of the individual. With his problematic identification thesis, that is to say, Benjamin construes the cinema as a medium in which that super-individual perfection of bodily life is approached which according to his psycho-physical schemata coincides with apocalypse (GS VI, 80; SW I, 395). The identification thesis is therefore an integral part of Benjamin's religious intentions. Koch observes that the thesis is 'dubious' since it disregards the possibility that the audience identifies with the apparatus not on a critical, Brechtian level but 'on the narcissistic level of an enormous extension of the perceptual apparatus'.[86] This criticism, however, concerns not a particular thesis but Benjaminian thought as a whole. For the clairvoyant and Scipio, whose 'true practice' we are to regain, may well be narcissists in their own right. And as argued in the present study, Benjaminian *Kritik*, including the esoteric dimension of its adaptation of Epic Theatre, itself conforms to mystical exercises (Sections 2.3.2 and 3.4.2). Instead of disregarding the possibility of 'narcissism', the identification thesis may actually combine the implicit affirmation of 'narcissism' with explicit appeals to critical reason.

To the extent that the thesis reflects actual developments in modern society, it concerns suffering and the risks of lasting alienation. Any technicalisation of the human body mortifies it and may result in an adaptation to those necessities that, as Benjamin of course knows, paralyse our sense of freedom (GS I.2, 499). Yet our sense of freedom is not an unequivocal good where revelation is being desired. For apocalypse to occur, the perfection of the 'body' of mankind must coincide with the *Untergang* of bodily life. The *corpus* in which man belongs to God will only be resurrected when the mortification of the human *Leib* is carried through to the very end. Then the body of mankind may have been salvaged '*in*' the mortification of *Leib* as the work of art, according to Benjamin, is salvaged '*in*' Baroque allegory or critical 'mortification' (GS I.1, 358). The possibility of such salvation at any rate explains why Mickey Mouse, for Benjamin, is not necessarily a dangerous illusion, but a possible vehicle of redemption ('erlösend').[87] By representing the actually mortifying unification of body and *Technik* as enjoyable, Mickey Mouse keeps us going in the direction of that destruction from which alone redemption might spring.

3. If the modern advent of apocalypse coincides with the perfect union of man and nature in *Technik*, any spectator who tests the artistic performance of the film actor for its adaptation to the requirements of cinematic technology resembles the Benjaminian critic who judges art on the basis of its revelatory potential. Any spectator whose degree of adaptation is tested in the cinema, in turn, resembles the critical writer whose own art and writing is a test that demonstrates the degree of his readiness to receive the gift of revelation. In section 1.3.3, I argued that in 'On Language' Benjamin aims to provoke revelation and attributes to language as such the fulfilment of this aim. The view that in 'The Work of Art' the same happens with regard to cinema explains the fact that Benjamin does not even address the psychological implausibility of his picture of a shocked and examined examiner. If the revelation to be

provoked occurs in a sphere of complete neutrality with regard to the subject–object opposition, this opposition and its psychological corollaries are not important in the *Darstellung* of the conditions of revelation in general, and the technicalisation of the human body in particular.[88] Whoever accepts that the examined examiner in 'The Work of Art' identifies with that cinematic apparatus of production that is communicated to him in the cinema, comes as close to accepting that cinema communicates itself in itself as Benjamin, in 'On Language', comes to the positive, but also suspended, affirmation of (L).

In the 'The Work of Art', then, Benjamin deals with secular art but does so in response to his desire for religion. In some of its main features, his conception of the decline of 'aura' agrees with this result. It can be summarised as follows: Traditional art drew its auratic qualities from its 'cult value' within pre-modern magical practices; these practices aimed for the domination of nature. In twentieth-century art, in contrast, there is a dominance of the 'exposition value' that is achieved by a combination of progressive technique and modern technology, the true *telos* of which lies outside of instrumental reason. Cultic art celebrates the mystery and relies on remoteness; expositional art exposes things and brings them close.[89]

This story is unconvincing. Since the eighteenth century, any auratic quality of art has rested at least as much on its emancipation from the realm of instrumental reason as on its magical heritage. Benjamin concedes this indirectly by calling the doctrine of *l'art pour l'art* a 'theology of art' (GS I.2, 481). As argued in section 1.2.2, the notion of magic in 'On Language' binds 'magic' to the alleged immediacy of spiritual communication and disregards those elements in magical practice that conform to the means-ends relation. In 1936, the non-instrumental character of true language has found a home in *Technik*, while 'magic' is reserved for practices that conform to the means-ends relation. Expositional art thus takes the place of what already the Benjamin of 1916 considered the language of *Technik* (and art), namely the revelatory communication of spiritual contents. The emphasis, in 1936, on closeness and 'exposition' at the expense of remoteness and depth is an integral part of his writing for revelation. In 'On Language', 'revelation' is opposed to 'the inexpressible'. According to the *Trauerspiel* study, 'deep' spirit ('der tiefe Geist') belongs to Satan and not to God (GS I.1, 406). The presentation of the Baroque as a secular age, moreover, dominates that study while the explicit expression of religious hope is relegated to the last few pages. We can think of 'The Work of Art' as a text in which the direct expression of religious hope has moved beyond the last of its written words. The reception of film and twentieth century art is described, to use Adorno's phrase, as they *would* appear from the standpoint of redemption,[90] but this standpoint is no longer explicitly referred to in the description. Before reformulating his messianism in 'On the Concept of History', a text not intended for publication, Benjamin entrusted the hope for redemption exclusively to his *Darstellung* of the profane.

3.5 The invincible dwarf

Benjamin described Reflection XVII of 'On the Concept of History' as a methodological summary of his former works.[91] In conclusion, I shall try to show that this description adequately captures his rhetoric of self-reference. Reflection XVII opposes 'materialist' historiography to an allegedly atheoretical historicism. The 'theoretical' or constructive principle of 'materialism' is what Benjamin calls 'Stillstellung' ('arrest'), namely the creation of a standstill in both thoughts and events.

> Zum Denken gehört nicht nur die Bewegung der Gedanken sondern ebenso ihre Stillstellung. Wo das Denken in einer von Spannungen gesättigten Konstellation plötzlich einhält, da erteilt es derselben einen Chock, durch den es sich als Monade kristallisiert. Der historische Materialist geht an einen geschichtlichen Gegenstand einzig und allein da heran, wo er ihm als Monade entgegentritt. In dieser Struktur erkennt er das Zeichen einer messianischen Stillstellung des Geschehens, anders gesagt einer revolutionären Chance im Kampfe für die unterdrückte Vergangenheit. [GS I.2, 702–3]

> It is not only the movement of thoughts, which is part of thinking but also their *Stillstellung*. Where thinking suddenly stops in a configuration saturated with tensions, it communicates a shock to that configuration, by which thinking crystallises into a monad. The historical materialist approaches a historical object only where it comes to him as a monad. In this structure he recognises the sign of a messianic *Stillstellung* of the course of events, or, put differently, of a revolutionary chance in the fight for the oppressed past.

Three things are important here: first, the *Stillstellung* of thoughts; second, the historical object, or subject matter, of the 'materialist', and, third, a *Stillstellung* of the course of events.

1. In the *Stillstellung* of thoughts, thinking communicates a shock to a configuration saturated with tensions. At first glance, thinking seems to play the active part of communicating the shock, and the configuration in question seems to be the object, or recipient, of the shock. Yet the distinctions between 'active' and 'passive', subject and object, or between 'shocking' thinking and a 'shocked' configuration thought are at odds with what Benjamin tries to say. First, with reference to the halt of thinking he uses not the verb 'aufhören' ('to cease') but 'einhalten', the meaning of which corresponds to a reflexive use of the English 'to arrest'. The runner who 'einhält' in his run does not merely stop running but 'arrests himself'.[92] A thinking which 'einhält' is also itself the object of its action. Second, what Benjamin calls configuration is not only the recipient of a shock, or the object of thinking, but also its medium; for thinking is said to stop *in* the configuration. Third, the sudden (self-)arrest of thinking in the configuration itself resembles a shock which *thinking* receives.[93] Benjamin concludes accordingly that thinking crystallises into a monad. Yet he concludes that thinking thus crystallises not by virtue of the shock that it receives, but by virtue of the shock that it communicates to the configuration. In his French translation, it is even the configuration that is said to undergo the crystallisation in question, and not thinking (GS I.3, 1265). And, finally, since 'crystallisation' denotes particular kinds of fixation in saturated solutions, Benjamin's image of the crystallisation into a monad suggests a 'shock' in the saturated 'configuration'. By actually using it for thinking, however, he subverts the difference between 'thinking' and 'configuration'.

In so far as the *Stillstellung* of thoughts is concerned, then, the constructive principle of 'materialist' historiography agrees with Benjamin's earlier postulates of self-reference in discarding the distinction between subject, object and medium. It differs from the postulate of self-reflection in his doctoral thesis by conceiving 'thinking', in accordance with his 'anthropological' interests, in the light of the psycho-physiological, or mystical, implications contained in 'shock'. The recourse to 'saturation' and 'crystallisation' recalls the chemical understanding of *Darstellung* as preparation, which Benjamin found in Schlegel. In *The Concept of Art Criticism*, it is the prosaic core in each work which criticism 'prepares' from and '*in*' the work's own reflection (GS I.1, 109; SW I, 178). In 'On the Concept of History', 'materialist' historiography so far seems to 'prepare' the monad '*in*' the (self-)arrest of thinking. As we shall see, however, Benjamin continues to consider the monad in question the core of a work.

2. The structure of a monad pertains not only to thinking but also to any historical object that the 'historical materialist' approaches. Unlike Benjamin's historicist, his 'materialist' does not deal with an amorphous mass of data but selects objects that present themselves as monads. Given that thinking has just been said to crystallise into a monad, one may suspect the monadological structure of the historical object to result from a projection of the historian's thinking onto its object. Benjamin tries to dispel that suspicion with his formulation that the object comes to the 'materialist' as a monad. The object, he insists, displays of its own accord the very structure into which self-related thinking crystallises; thinking and object are both 'monads', things of the same kind. Historical knowledge, as conceived in 'On the Concept of History', therefore agrees with what Benjamin in his thesis discussed as the basic assumption of Romantic epistemology, namely the assumption that any knowledge of an object presupposes the self-knowledge, or 'reflection', of that object (Section 2.2.3). In 'historical materialism', just as in the 'reflection' of *The Concept of Art Criticism*, we find not a subject's knowledge of an object, but the mutual transition between, or communion of, the object and subject.

3. The final sentence of the passage under discussion can strengthen the suspicion of projection. The monadological structure of the historian's object, the sentence implies, is not necessarily the structure of historical events. In the monad as which his object 'comes' to him, the 'materialist' recognises not a *Stillstellung* of the course of events, which would correspond to the *Stillstellung* of his thoughts, but merely the sign of a factual *Stillstellung*. The historical object that comes as a monad, that is to say, points to a *Stillstellung* that may or may not agree with the underlying structure of historical reality. The 'historical materialist', by consequence, may well encounter a 'sign' that promises, but will never correspond to, the fulfilment of his desire. Only if the messianic *Stillstellung* of the course of events occurs, will the apocalyptic truth have been anticipated in the historian's communion with his monadological object. Like Benjamin's earlier postulates of self-reference, the alleged crystallisation of self-related thinking in 'historical materialism' will only prove to be knowledge at the end of time.

To the monad of Reflection XVII pertains that intensive infinity which Benjamin in 'On Language' attributed to self-communicating language and, in *The Concept of Art Criticism*, to the reflection in literature and criticism. 'On the Concept of History' joins the *Trauerspiel* study in using Leibniz's concept to illustrate the infinity in question, and also in offering a more obviously 'historical' version of self-reference than the thesis (Section 3.2.1). The 'historical materialist', we read, seizes upon the revolutionary chance in the monadological structure of his object in order to blast a specific era out of the homogeneous course of history, or a life out of its era, or an individual work out of its author's *œuvre*. In his apparent fixation on individual 'monads', he *appears* to lose himself in the representation and characterisation of individual things. In this appearance, his thinking resembles Schlegel's 'romantic *Poesie*' (KA II, 182–3). With its destruction of continuity and context, 'materialism' also resembles romantic *Poesie* in striving for the disclosure of a higher nexus or *Zusammenhang* in which everything – 'the entire surrounding world' in Schlegel's words, 'the entire course of history' in Benjamin's – is contained in the intensive infinity of the one, self-enclosed object. What the method of Benjamin's materialist 'yields' is 'that *in* the individual work the *œuvre* is preserved and sublated, *in* the *œuvre* the era, and *in* the era the entire course of history'.

The different levels of historical study which Benjamin distinguishes here all find their counterparts in the ever more general frames of reference in which Schlegel situates critical understanding.[94] Schlegel holds that the understanding of any particular work requires the understanding of its place in that system which is formed by all the works of the author in question; Benjamin has his historical materialist find the *œuvre 'in'* the individual work. Schlegel does not explicitly discuss the understanding of a period, yet his demand that the work be related to congenial ones which may be centuries apart certainly fulfils Benjamin's requirement to 'blast' the work, or the *œuvre*, out of its era. And the 'entire course of history' which Benjamin's materialist recognises *'in'* the era – and thus *'in'* the *œuvre* and *'in'* the individual work as well – finds its counterpart in the so-called organism of all arts and sciences, which Schlegel seeks in human culture and history as a whole. The 'materialist' explosion of historical continuity envisaged by Benjamin thus actually conforms to what he in his thesis considered the central intention of Romantic criticism, namely the perfection, completion and systematisation – or the 'destruction' – of the work.[95]

For Schlegel, the organism of all arts and sciences tends to be indistinguishable from its own history, from the universal history, that is, of human *Bildung* and culture (KA II, 411). Like historicism, Romantic criticism strives for historical universality. And so does Benjamin's 'historical materialism': 'It is not the case that every universal history has to be reactionary. A universal history w i t h o u t a constructive principle is. The constructive principle of universal history [...] is a monadological one. It exists in *Heilsgeschichte*' (GS I.3, 1234). Neither an abandonment of universality nor the absence of religious intentions distinguishes Benjamin's 'historical materialist' from his historicist opponents.[96] The constructivist vocabulary of Reflection XVII can suggest that the decisive difference lies in the materialist's rejection of organic conceptions of history. Yet in the last sentence, Benjamin complements his chemical imagery by the very biological one that in *The Concept of Art Criticism* served to illustrate the critical dissolution into the absolute. He compares time to a seed that is contained in the 'fruit' of historical comprehension (or its object): 'The nourishing fruit of what has been historically grasped has time as the [...] seed *within* it.' The 'seed' of time in the object of the 'historical materialist' corresponds to the 'germ' of reflection in that of the romantic critic. The time of the 'materialist', we may say, is a full and infinite medium ('eine mediale und qualitative [Unendlichkeit]') (GS I.1, 92; SW I, 168). For Benjamin, the parallels between 'literary self-reflection' and 'historical materialism' were of substantive importance. Of prose ('Prosa'), which in the thesis figured as the idea of *Poesie*, he states in one of his latest notes that its idea 'coincides with the messianic idea of universal history'.[97]

The parallels between the postulate of literary self-reflection and 'historical materialism' suggest that the 'material' life of past generations can fruitfully be compared to a text. What Benjamin considers the 'true' historian's 'philological' method makes sense only to the extent that life resembles the 'Book of Life' (GS I.3, 1238). In 'On the Concept of History', 'historical materialism' designates the attempt to recreate history in the likeness of the author's literary vocation.[98] The *writing* of history, accordingly, assumes an enormous significance. In Reflection XVI, we read: 'The historical materialist cannot part with the notion of a present which is not a transition, but in which time has stopped and come to a standstill. For that notion defines precisely that present in which he, for his part, is writing history' (GS I.2, 702). Read as a comment on the tradition of Marxist thought, this passage agrees with those of Benjamin's notes that advocate a religious re-interpretation of Marx's concept of classless society (GS I.3, 1231–2). However, Benjamin here is not primarily concerned with the indispensability of a messianic notion of time to political thought, but with the indispensability of that notion to a particular way of writing. The 'historical materialist' allegedly writes history *in* a present in which time 'has stopped',[99] in which, that is, the messianic arrest of time has already occurred. This claim neglects the difference between the messianic *Stillstellung* of time on the one hand and the sign of that *Stillstellung*, which the 'materialist' finds in the structure of the monad, on the other. The neglected difference, of course, is not a minor one; it may well be indispensable to ascertain the truth (or falsity) of 'historical materialism' in any secular sense of 'truth'. But the difference is only decisive where religion (or magic) has no say. Whoever accepts a duality similar to that which Benjamin accepted in 1919, a duality, that is, between the 'external side' of the world and its 'inner essence', or between the realm of practical action or 'works' on the one hand, and the 'orders of prayer' on the other, will tend to disregard the difference in question.[100] For he believes that there is a reality in the sign as such, the 'essential' reality, moreover, compared to that of the 'external side of the world'. The claim that the 'historical materialist' writes 'in' the messianic cessation of time similarly implies that reality is found in the mystic's *nunc stans*, and not exclusively in the 'external side' of, say, the Second World War. The claim, that is to say, agrees with Benjamin's postulate of linguistic self-communication in implicitly 'anticipating' the fulfilment of apocalyptic desire.

In contrast to the apodictic introduction of (L) in 'On Language', however, Benjamin in Reflection XVI of 'On the Concept of History' indicates the subjective character of the picture which he presents. The messianic present is the medium in which the 'materialist' writes history, but it is that only 'for his part' ('für seine Person'). As far as the 'materialist' is concerned, his writing already occurs in the messianic standstill of time, but the 'materialist' may be wrong. Accordingly, the author Benjamin distinguishes elsewhere between the messianic *Stillstellung* and the sign of it found in 'materialist' thought (or its object), or between the present ('Jetztzeit') as a model of messianic time and that time itself (GS I.2, 703). His attribution of 'a *weak* messianic power' to his generation, moreover, is conditional. It rests on the condition that certain impressions on which the historical materialist relies in his representation ('Vorstellung') of the past are correct (GS I.2, 693–4). In Benjamin's notes, the materialist's remembrance ('Eingedenken') even figures as the proverbial straw at which the drowning man clutches. It figures as an option, that is, which is unlikely to succeed (GS I.3, 1243–4). On the one hand, then, we ought not to identify Benjamin with the persona of historical materialist which he creates to express some of his intentions. The fact that the 'materialist' believes that he is writing history in the messianic standstill of time does not mean that Benjamin, for his part, believes that he is doing the same. Not even in his last rhetorical exercise of 'self-reference' has Benjamin parted with the religious scepticism of his youth. On the other hand, the very creation of a persona that finds in its writing messianic fulfilment is only comprehensible if religion is being desired. And the German phrase 'er für seine Person' may do more than merely qualify the placement of 'materialist' historiography in messianic fulfilment as a personal belief. In the given context, it *may* indicate that the historical materialist writes history for the sake of his person or personal salvation: 'er [schreibt] für seine Person Geschichte'. Since, moreover, the phrase is frequently used to express one's resolution to exempt a view or decision from (further) discussion, Benjamin may also be implying that as far as his 'materialist' is concerned, materialist historiography is situated in the standstill of time, period. Reflection XVI does not require us to ascribe such a decisionist gesture to Benjamin's 'materialist' (or his author). On the whole, however, 'On the Concept of History' can be described as an expression of religious decisionism. This is what Reflection I suggests in conjunction with Benjamin's notes.

In Reflection I, Benjamin does not characterise the thoughts, beliefs and intentions of his 'materialist', but comments on the relation between 'historical materialism' and 'theology'. Reflexion I thus provides the author's interpretation of a relation which is a central problem in 'On the Concept of History' and which, as an example of the problematic relation between religion and the various 'theories' that Benjamin proposes in his writings, also concerns his *œuvre* as a whole. The reflection starts with the famous description of a chess automaton that guarantees a winning game by permitting an invisible hunch-backed dwarf, who happens to be an unbeatable player, to direct the moves of a puppet in Turkish attire. Benjamin adds the following comment: 'One can imagine a counterpart of this automaton in philosophy. The puppet which is called 'historical materialism' is to win all the time. It can easily be a match for anyone if it enlists the services of theology, which today, as everybody knows, is small and ugly, and in any case has to keep out of sight' (GS I.2, 693).

As Pfotenhauer points out,[101] Benjamin's comment changes the example; in the example, the dwarf uses the puppet, in Benjamin's comment, the puppet enlists the services of the dwarf. Moreover, it is of course questionable whether we can imagine a *philosophical* collaboration between theology and Marxism. These problems, however, result less from Benjamin's text than from the attempt to use it for the purposes of theoretical discussion. To the extent that 'historical materialism' designates the application to history of a religiously conceived body-oriented anthropology, Benjamin's 'materialist' becomes indistinguishable from a theological historiographer. The question who uses whom, or who has priority over whom, is hardly fruitful where the candidates in question tend to merge. A 'historical materialist' in the indicated sense, furthermore, is not a historical materialist in Marx's sense even if the former happens to wave a red flag (or if Marx, against his intentions, should have developed a religious conception of social change). Theology can be theoretically compatible with any religiously defined 'historical materialism' whereas its basic assumptions are theoretically incompatible with Marxism to the extent that Marxism is adequately understood as a secular theory (with or without historical roots in religion).

The intriguing problem of Benjamin's comment lies in the fact that the realm in which he finds the counterpart of the chess automaton is imaginary. At least since the emancipation of philosophy from theology, the theologian has not had the winning cards in the philosophical game. Albeit not excluding him from philosophical discussion, his assumption that religious hopes will eventually be – or partially already are – fulfilled does not in itself represent an advantage over other players. We can imagine a counterpart of Benjamin's automaton in the realm of philosophy only if we take philosophy to be an enterprise in which success is measured by the degree to which religious hopes can be accommodated. The game, that is to say, in which Benjamin's 'historical materialist' is to win all the time is situated in what the author, already in 1917, called the 'coming' philosophy. Benjamin, who devoted much of his life's work to the realisation of the 'virtual' union of religion and philosophy, at no point was unrealistic enough to consider that realisation accomplished. In the 'today' of 'On the Concept of History', no open or liberating unification of philosophy and religion has materialised; it is still the ugly dwarf theology on whose hidden services the 'materialist' must rely. The non-existence or social irrelevance of that realm in which the dwarf and his puppet are unbeatable makes the imagination of power which is part of Benjamin's comment indistinguishable from a bitter acknowledgement of impotence. There is little comfort in possessing the winning strategy in a game the rules of which still have to be revealed, established or recognised. And, of course, the fact that actually the imagined game is not played in modern philosophy is of minor importance compared to the fact that it was also absent from the 1940 theatre of political and military action.

Notes

Introduction

1. 'A German Institute of Free Research', GS III, 525–6.
2. 'Wissenschaft als Beruf', pp. 588–613 (trans. pp. 134–56).
3. For philosophical discussions of 'self-deception', see David Pears, *Motivated Irrationality*; and Brian P. McLaughlin, Amélie Oksenberg Rorty (eds), *Perspectives on Self-Deception*.
4. For recent attempts at simultaneously affirming religion and discarding positive belief, see for instance Eric L. Santner, *On the Psychotheology of Everyday Life*; and Don Cupitt, *After God*. For a sociological analysis of the vicissitudes of modern religion and religiosity, see Steve Bruce, *Religion in the Modern World*.

Chapter 1

1. Cf. Hamann, *Aesthetica in nuce*, p. 198.
2. Cf. Johannes Micraelius, 'communicatio realis'; and Micraelius, 'communicatio verbalis'.
3. See 'Mittheilen', DW VI; and Jürgen Barkhoff, *Magnetische Fiktionen*.
4. Emil Lask, *Die Logik der Philosophie und die Kategorienlehre*, pp. 27–44, 58–73; Martin Heidegger, *Die Kategorien- und Bedeutungslehre des Duns Scotus* (actually a study of Thomas von Erfurt), pp. 238–9, 264, 405–6.
5. Winfried Menninghaus, *Walter Benjamins Theorie der Sprachmagie*, pp. 11–13. Unlike Menninghaus, I consider Humboldt a possible source for Benjamin's notion of spiritual self-expression but not for his notion of linguistic self-communication. Cf. the criticism of Humboldt in 'Reflections on Humboldt' (1925–28), GS VI, 26–7, SW I, 424.
6. Menninghaus, *Sprachmagie*, pp. 13–15.
7. GS II.1, 161, SW I, 103 (*Coming Philosophy*).
8. Benjamin's use of 'Wesen' is one reason for the co-existence of subject-oriented and objective, metaphysical interpretations of 'On Language', *and* can seem to support the deconstructive rejection of both. See, for instance, Regine Kather, *'Über Sprache überhaupt und über die Sprache des Menschen'*, p. 21; Thomas Schwarz Wentzer, *Bewahrung der Geschichte*, pp. 103–4; and Bettine Menke, *Sprachfiguren*, pp. 9–11, 37, 48–51.
9. For Benjamin's early writings on painting, see Uwe Steiner, *Die Geburt der Kritik aus dem Geiste der Kunst*, pp. 46–90; and Howard Caygill, *Walter Benjamin: The Colour of Experience*, pp. ix–23. In 'The Rainbow: A Dialogue on Imagination' (1914–16), Benjamin conceives 'pure' seeing as self-referential, as one dimension, that is, of what in 'On Language' he calls 'language'. GS VII.1, 23; and Letter to Gershom Scholem, 22 October 1917, GB I, 395.

10. After the introduction of (L), Benjamin states that given the truth of (L), the distinction between spiritual and linguistic essence 'turns into' – rather than simply is – a tautology (145; *66*).
11. In 'Programme of the Coming Philosophy' (GS II.1, 157–71, SW I, 100–10) Benjamin conceives 'experience' in such a way that 'knowledge' and 'religion' merge. Any account of his notion of experience which considers it philosophical *and* tends to disregard his interest in religion is therefore problematic. See the related discussions in Susanne Heil, *'Gefährliche Beziehungen'*, pp. 61–3; and Caygill, *Colour*, pp. 1–7, 23.
12. *Wissenschaft der Logik* (vol. XI), pp. 351–3; trans. (vol. I), pp. 139–42.
13. For Benjamin's awareness of the fact that self-reference is a possible source of the nonsensical, see his notes on Russell's paradox (GS VI, 9–11). Cf. Alfred, North Whitehead and Bertrand Russell, *Principia Mathematica*, esp. vol. I, pp. 37–65; and Caygill, *Colour*, pp. 14–15.
14. For different assessments see Menninghaus, *Sprachmagie*, esp. p. 17; Schwarz Wentzer, *Bewahrung*, pp. 124, 113; and Menke, *Sprachfiguren*, pp. 39, 48–53.
15. Benjamin's use of 'magic' in 'Reflections on Humboldt' and 'Doctrine of the Similar' (1933) is equally guarded. See GS VI, 26, SW I, 424, GS II.1, 208, SW II, 697; and Menke, *Sprachfiguren*, pp. 31–2.
16. In his letter to Martin Buber of 17 July 1916 (GB I, 325–8) Benjamin uses 'magic' to advocate the ideal of contemplation in the face of both the First World War and the addressee's pro-war stance. (Cf. Anson Rabinbach, 'Between Enlightenment and Apocalypse', p. 105.) The image of a magical spark that flashes over from 'immediate' word to effective deed, however, also suggests that the advocated contemplation represents a higher form of action. For Benjamin's related understanding of prayer see section 1.3.2.
17. I analyse Benjamin's introduction to his interpretation of Genesis, which contains the claim that language is an inexplicable reality, in section 1.3.2.
18. Cohen, *Religion der Vernunft aus den Quellen des Judentums*, pp. 55–6; Husserl, *Philosophie als strenge Wissenschaft*, pp. 32–7, 61 (trans. pp. 110–16, 146).
19. See 'Dialogue on the Religiosity of the Present' (1912), GS II.1, 16, and section 2.1.
20. Hans Robert Jauß, *Kleine Apologie der ästhetischen Erfahrung*, Jauß, *Ästhetische Erfahrung und literarische Hermeneutik*; Rüdiger Bubner, *Ästhetische Erfahrung*, pp. 52–69; Jens Kulenkampff, 'Ästhetische Erfahrung – oder was von ihr zu halten ist'; and George Dickie, 'The Myth of the Aesthetic Attitude'. For discussions of Benjamin's work which focus on 'aesthetic experience', see Helmut Pfotenhauer, *Ästhetische Erfahrung und gesellschaftliches System*; Alexander Pivecka, *Die künstliche Natur*; and Caygill, *Colour*.
21. 'Bewußtmachende oder rettende Kritik', p. 200 (trans. p. 46).
22. Georg Büchner, *Dantons Tod*, II: 3, p. 43. For Benjamin's own disposition to adopt an attitude similar to Lucile's, see 'Moscow Diary' (1926–27), GS VI, 303.
23. John McCole, *Walter Benjamin and the Antinomies of Tradition*, pp. 2–10; Birgit Recki, *Aura und Autonomie*; Marleen Stoessel, *Aura*; and Rainer Rochlitz, *The Disenchantment of Art*, pp. 149–66, 207–20.
24. Stoessel, *Aura*, p. 157.
25. According to a related note, it is not the content but the essence of language which defines its boundaries (GS VII.2, 786).

26. *Summa Theologiæ*, I, 4, 2, c. For what follows, see S. theol. I, 1–4.
27. Saint Thomas Aquinas, *L'Être et L'Essence*, editor's introduction, pp. 8–9.
28. *S. Thomae Aquinatis In Librum de Causis Expositio*, 23, n. 387.
29. Cf. Benjamin's use of 'absolute' in 'On Painting, or Sign and Mark' (1917–18), GS II.2, 605, SW I, 84.
30. (L) approaches the status of a definition of language since it concerns all language and is closely related to what Benjamin considers the essence of language: name.
31. *A Rhetoric of Motives*, pp. 275–6, 298–301.
32. Cf. GS I.1, 216–7, GS VI, 25–6; and Hermann Güntert, *Von der Sprache der Götter und Geister*, p. 49. For Benjamin's corresponding use of Hermann User's *Götternamen*, see Rudolf Speth, *Wahrheit und Ästhetik*, p. 260.
33. Cf. Hans Ludwig Held, 'Über Golem und Schem', p. 359. Benjamin was familiar with at least the first part of this study (GB I, 344).
34. Benjamin's interpretation of the genitive in 'the language of language' as concerning language as a medium, and not as a means, suggests that 'medium' implies the identity of the subject and the medium of communication. Given that in his treatment of human language he considers as identical what he had introduced as the subject and the object of communication respectively, namely spiritual and linguistic essence, this understanding of 'medium' does not invalidate the interpretation of the term proposed in section 1.2.2. For a related use of 'medium' in a fragment of 1920, see GS VI, 23.
35. See the related notes, GS VI, 11–14, SW I, 87–9.
36. For the enigmatic character of 'name' in Benjamin's published writings, see Michael W. Jennings, *Dialectical Images*, p. 155.
37. Cf. Scholem, *The Messianic Idea in Judaism and Other Essays on Jewish Spirituality*, pp. 288–303; and Scholem, 'The Name of God', p. 180.
38. Cf. Menke, *Sprachfiguren*, p. 54. As the condensation of a wide-ranging, 'systematical', assumption, 'name' is an example of what Benjamin discusses in his doctoral thesis as 'mystic terminology'. See GS I.1, 47–50, SW I, 139–42, and section 2.3.2.
39. The German word 'Reflex' denotes both a reflex action and a reflected image or the reflection underlying it.
40. In section 1.3.3, I shall argue that it is more appropriate to consider 'On Language' not a theoretical and theological contribution but a piece of post-traditional, religious prose. The term 'theory' is sometimes used to distinguish Benjamin's thinking from 'philosophy' and discursive argument. (See for instance Sigrid Weigel, *Body- and Image-Space*, pp. 12, 58.) The corresponding view of Benjamin as a 'theorist' is compatible with my understanding of his writing as essentially atheoretical.
41. Terry Eagleton, *Walter Benjamin or Towards a Revolutionary Criticism*, p. 19; Menninghaus, *Sprachmagie*, pp. 18, 21–2, 32; Schwarz Wentzer, *Bewahrung*, pp. 122, 130–3; Bernd Witte, *Walter Benjamin*, p. 28.
42. For a different assessment, see Jennings, *Dialectical Images*, pp. 90, 96. Norbert Bolz and Willem van Reijen (*Walter Benjamin*, pp. 31–40) adopt Adorno's notion of 'inverse theology' (AB, 90) to characterise Benjamin's stance. It is not clear in what an inverse theology in the evoked sense differs from a theology which is both uninverted and concealed.

43. 'The Name of God', esp. p. 70. See also Susan A. Handelman, *fragments of Redemption*, pp. 1–115 and, for the view that Benjamin 'may be considered one of Scholem's predominant sources', David Biale, *Gershom Scholem*, p. 103.
44. Cf. Letter to Ludwig Strauß, 11 September 1912, GB I, 64.
45. The term 'secular' is sometimes used to characterise a revival of religious preoccupations outside of traditional religious communities or institutions. (See, for instance, Richard Wolin, *Labyrinths*, p. 54.) I do not question the relevance of such a weak notion of the secular to Benjamin's *œuvre* but argue against the adequacy of 'secularisation' in the strong sense, according to which that which is secular or securalised is not, or is no longer, religious in nature.
46. *Athenäum* fragment 116, KA II, 182. See also section 2.2.3.
47. For a likewise paradoxical treatment of 'identity', see 'Theses on the Problem of Identity' (1916), GS VI, 27–9, SW I, 75–7.
48. Like my presentation in general, the modernising translation of 'ratio' by 'sense' simplifies the issues involved in the medieval debates on analogical predication. For introductions to *analogia entis*, more detailed and less sceptical than the account given above, see John F. Wippel, 'Metaphysics'; and Josef de Vries, *Grundbegriffe der Scholastik*, pp. 25–37.
49. S. theol. I, 13, 5 c.
50. de Vries, *Grundbegriffe*, p. 33.
51. Cf. S. theol. I, 13, 5 c.
52. Cf. Aquinas's wording 'non [...] ex toto est eadem ratio' in his *De prinicipiis naturae*, para. 6.
53. S. theol. I, 19, 2 c.
54. 11 November 1916, GB I, 343.
55. Moses Maimonides, *Guide for the Perplexed*, I, 56, and S. theol. I, 13.
56. See, also for what follows, GS II.1, 158–66, SW I, 100–7 ('Coming Philosophy'); and Heidegger, *Kategorien- und Bedeutungslehre*, pp. 255–63, 402–10.
57. See, also for what follows, Lask, *Logik der Philosophie*, pp. 135, 223–43, 268; and Speth, *Wahrheit*, pp. 27–31. Astrid Deuber-Mankowsky (*Der frühe Walter Benjamin und Hermann Cohen*, pp. 38–50) discusses the importance to Benjamin's response to neo-Kantianism of Felix Noeggerath's related ideas.
58. For Benjamin's reception of Heidegger's *Habilitationsschrift*, see GB II, 68, 108, 127; and GS VI, 642–5.
59. See GS VI, esp. 9–40, 639–55 and GS VII.2, 785–90. Benjamin's terminological inconsistencies and experiments may reflect his attitude towards the concept of revelation, which I discuss in section 1.3.2.
60. GB I, 326–7, GS VI, 18, SW I, 268, GS VII.2, 786; Witte, *Walter Benjamin: Der Intellektuelle als Kritiker*, pp. 69–79; and Speth, *Wahrheit*, pp. 200–8.
61. For a different view see Menke, *Sprachfiguren*, p. 49.
62. *Phänomenologie des Geistes*, pp. 122–3 (trans. p. 127).
63. 'Philologische Einfälle und Zweifel über eine akademische Preisschrift', pp. 46–8.
64. Hamann, *Briefwechsel*, V, p. 177. The passage quoted above will provide Heidegger with the starting-point for a meditation on self-contained

language which he, too, experiences as the hovering above an abyss. *Unterwegs zur Sprache*, p. 13 (*Poetry, Language, Thought*, p. 191).

65. See also Menke, *Sprachfiguren*, pp. 48–54; and 'On Riddle and Mystery' (GS VI, 17–8, SW I, 267–8). This fragment shows that the early Benjamin uses 'word' and 'symbolic' to justify the desire for mystery in the face of the (apparently) profane. This intention remains essential to what in the surrealism essay of 1929 he chooses to call 'profane illumination'. See section 3.3.2.
66. For the historical context of Benjamin's life-long concern with *Ursprung* ('origin'), see Wolin, *Walter Benjamin*, pp. xix-lii; and Christopher J. Thornhill, *Walter Benjamin and Karl Kraus*.
67. Cf. Letter to Scholem. 22 October 1917, GB I, 391–2. For the considerable divergence of Benjamin's understanding of 'revelation' from Judaism, see Gillian Rose, 'Walter Benjamin – Out of the Sources of Modern Judaism', pp. 184–8.
68. For an important number of Romantic readers of the Bible, the poetic and religious dimensions of the biblical text tend to become indistiguishable. (Stephen Prickett, *Words and 'The Word'*, pp. 37–68.) This is precisely a reason, however, for not considering such readers believers in divine revelation.
69. 146; *66*. For a biographical account of Benjamin's proximity to the circle around George, see Momme Brodersen, *Walter Benjamin*, pp. 109–18, 124–32. Speth (*Wahrheit*, pp. 34–7) discusses the influence of Norbert von Hellingrath on the notion of *Gestalt* in Benjamin's Hölderlin essay. For Benjamin's understanding of *Gestalt* within the religious framework of his early anthropological notes, see GS VI, 78–80, SW I, 393–5.
70. 151; *70*. Hamann, 'Des Ritters von Rosencreuz letzte Willensmeynung', p. 32. In 'The Task of the Translator' (1923) Benjamin refers directly to John 1: 1 in order to support his ideal of translation (GS IV.1, 18, SW I, 260). Were it not for the religious implications in his metaphysical notion of translation, this use of John 1: 1 could be considered an outright blasphemy.
71. For the view that 'secularisation' is itself a religious concept, see Hans Blumenberg, *Säkularisierung und Selbstbehauptung*, pp. 9–18, 75–89.
72. *Sprachmagie*, pp. 7–8.
73. Stoessel, *Aura*, pp. 83–5.
74. For a positive assessment of Benjamin's distance from both philosophical and philological argument, see Adorno, 'Einleitung zu Benjamins *Schriften*', pp. 573–4; and Adorno, 'Charakteristik Walter Benjamins', pp. 240–1.
75. Benjamin's references, in 'The Task of the Translator', to sacred literature in general may confirm the point made above. GS IV.1, 21, SW I, 262–3.
76. At least in their imagery, Benjamin's comments on the Bible anticipate his characterisation, in *The Concept of Art Criticism*, of literary criticism as the unfolding of the 'germ' ('Keim') of reflection, present in the work being criticised (GS I.1, 73, 78, SW I, 156, 159). For a non-mystical interpretation of Benjamin's notion of the unfolding of an inexplicable reality see Menninghaus, *Sprachmagie*, pp. 33–4.
77. Benjamin certainly considers what he calls 'the facts of language' to be part

and parcel of the world's furniture, and John presents all reality, knowledge and life as originating in God or the (biblical) word. The interpretation of John 1: 1–4 as self-referential is in structural agreement with the equation of God's creative word with both the name of God and the mystic Torah, which Scholem ('The Name of God', p. 70) considers essential to cabbalistic thought.

78. Cf. Edmund Jephcott's translation 'and the Bible is only *initially* indispensable for this purpose'. SW I, 67.
79. See also GS I.1, 196, SW I, 351 ('Goethe's Elective Affinities') and GS I.1, 211 (*Trauerspiel* study).
80. *Zur Kabbala und ihrer Symbolik*, pp. 63–4; 'The Name of God', p. 175.
81. For a discussion of the equation of the object with the medium of communication in traditional Christian belief see Jürgen Werbick, 'Das Medium ist die Botschaft'.
82. John Baillie, *The Idea of Revelation in Recent Thought*, pp. 3–40; Tobias Trappe, 'Selbstoffenbarung' (HWP IX); Werbick, 'Medium'.
83. *De signatura rerum*, pp. 193–5, 203–4.
84. Cf. Friedrich, Christoph Oetinger, *Die Lehrtafel der Prinzessin Antonia*, pp. 93–5, 149–51, 166–7.
85. Cf. Schelling, *Philosophische Untersuchung über das Wesen der menschlichen Freiheit*, pp. 338–41 (trans. pp. 73–7); Schelling, *System des transcendentalen Idealismus*, pp. 600–1 (trans. pp. 208–10); and Franz von Baader, *Vorlesungen über J. Böhme's Theologumena und Philosopheme*, pp. 387–90. For Benjamin's appreciation of Baader's work see Brodersen, *Walter Benjamin*, p. 83.
86. *Zur Dogmatik*, pp. 61–2.
87. *System der christlichen Glaubenslehre*, I, pp. 164–71, 448–59.
88. See Trappe, 'Selbstoffenbarung' (HWP IX) and the discussion in Werbick, 'Medium', pp. 203–15.
89. In both the Jewish and the Christian context, dismissals of the notion of God's self-revelation as relatively unimportant seem to be linked to the firm and unquestioned conviction that God exists. Cf. Moses Mendelssohn, *Jerusalem*, pp. 415–20; and Richard Swinburne, *Revelation*, pp. 2–5.
90. Cf. *Der Stern der Erlösung*, pp. 174–228, esp. p. 179 (trans. pp. 156–204, esp. pp. 160–1). For Benjamin's reception of Rosenzweig, see Stéphane Moses, 'Walter Benjamin and Franz Rosenzweig'.
91. For Benjamin's commitment to the so-called *Jugendkulturbewegung* ('youth culture movement'), see the documentation in GS II.3, 825–88; Brodersen, *Walter Benjamin*, pp. 23–76; Deuber-Mankowsky, *Benjamin und Cohen*, pp. 299–313; Michael Bröcker, *Die Grundlosigkeit der Wahrheit*, pp. 13–67; and McCole, *Antinomies*, pp. 35–70.
92. For the importance of the journal *Die Tat*, see Kurt Sontheimer, 'Der Tatkreis'. Sections 1.3.3 and 2.1 focus on 'The Religious Position' and 'Dialogue on the Religiosity of the Present', GS II.1, 72–4 and 16–35 respectively. Unless otherwise indicated, references in these sections are to these texts.
93. Letter to Strauß, 9 September 1912, GB I, 64.
94. GS II.1, 31. For Benjamin's use of 'Führer', see Ansgar Hillach, '"Ein neu entdecktes Lebensgesetz der Jugend"'.
95. See Søren Kierkegaard, *Either/Or*, II, p. 179; Speth, *Wahrheit*, pp. 167–72;

Sontheimer, *Antidemokratisches Denken in der Weimarer Republik*, pp. 327–31; Martin Greiffenhagen, *Das Dilemma des Konservatismus in Deutschland*, pp. 248–52, 268; and Thornhill, *Political Theory in Modern Germany*, pp. 67–91. In some instances, the decisionistic phrasing of 'The Religious Position' strongly resembles that in Buber's *Drei Reden über das Judentum* (pp. 27, 42, 52–3, 59–63; trans. pp. 19, 26, 31, 34–6), with which Benjamin was familiar (GB I, 71).

96. Cf. GS I.2, 440, 480 ('The Work of Art') and Benjamin's related discussion of closeness and distance in 'Schemata on the Psycho-Physical Problem' (1922–23), GS VI, 83–7, SW I, 398–400. See also Recki, *Aura und Autonomie*, pp. 26–9, 49–59 and, for the *topos* of 'remote closeness' in mystical writing, Dietmar Mieth, 'Die Fremdheit Gottes in Mystik und Moderne', pp. 107–10.
97. NE 1177a12–1179a32.
98. H. Nottarp, 'Gottesurteil: I. Rechtsgeschichtlich'; and E. Kutsch, 'Gottesurteil: II. In Israel'.
99. See also 'Critique of Violence' (1921), GS II.1, 202–3, SW I, 252.
100. See also sections 3.4.1–3.
101. Lytton Strachey, *Eminent Victorians*, p. 89.
102. It is the argument known as 'Pascal's Wager' which convinces Pascal's sceptic that the belief in God is desirable. See, also for what follows, Blaise Pascal, *Pensées*, pp. 550–1 (trans. pp. 92–6).
103. Weber, 'Wissenschaft als Beruf', pp. 588–613 (trans. pp. 134–56). See also Witte, *Der Intellektuelle*, pp. 149–58; Herbert Schnädelbach, *Philosophy in Germany*, pp. 21–32; Heil, *'Gefährliche Beziehungen'*, pp. 68–80; and Rabinbach, 'Enlightenment and Apocalypse', pp. 87, 101.
104. For a sympathetic but nevertheless devestating account of Cohen's fusion of Judaism and Kantianism as being in conflict with both Kantian rationality and essential elements of the Judaic tradition, see Rose, 'Hermann Cohen – Kant among the Prophets'.
105. In his *Aporien und Apologie*, Michael Rumpf criticises both Benjamin's writings and their reception as a variety of 'Erbauungsliteratur' (roughly, 'devotional literature'). The major limitation of his refreshing and valuable critique lies in the assumptions, first, that academic literature simply must not be what it widely – in some quarters, at least – is, namely 'devotional' in character, and second, that Benjamin's early writing is completely arbitrary. In reality, it is precisely the pragmatic aim of post-traditional edification, disliked by Rumpf, which provides these writings with their alogical consistency.

Chapter 2

1. Brodersen, *Walter Benjamin*, pp. 42–5.
2. Letter to Wyneken, 9 March 1915, GB I, 263–4.
3. In accordance with Early Romantic usage, Benjamin in his thesis uses 'Kunst' ('art') – the German umbrella term for the visual arts, literature and music – mainly in the narrower sense of 'literature' or 'Dichtung'. I therefore frequently speak of literature where Benjamin uses 'Kunst'.
4. Letters to Strauß, 11 September and 10 October 1912, GB I, esp. 64, 69–70.

5. See Wyneken, *Schule und Jugendkultur*, pp. 163–74; and Heinrich Kupffer, *Gustav Wyneken*, pp. 53–4.
6. Kupffer, *Gustav Wyneken*, p. 44.
7. Letter to Strauß, 10 October 1912, GB I, 71. See also Moritz Goldstein, 'German Jewry's Dilemma before 1914'; Rabinbach, *Enlightenment and Apocalypse*, pp. 89–99; Gary Smith '"Das Jüdische versteht sich von selbst"'; and Deuber-Mankowsky, *Benjamin und Cohen*, pp. 282–312.
8. Kupffer, *Gustav Wyneken*, pp. 168–73.
9. *Werke und Tage*, pp. 242–3, quoted in Werner Fuld, *Walter Benjamin*, p. 38.
10. 'Wandervogel und Freie Schulgemeinde', p. 135.
11. *Jugendkultur*, p. 181.
12. Cf. *Jugendkultur*, pp. 163–74.
13. For the influence, on Wyneken's thought, of Hans Vaihinger's post-Kantian philosophy of the 'as-if', see Kupffer, *Gustav Wyneken*, pp. 17, 209–12.
14. Cf. *Jugendkultur*, p. 162.
15. *Jugendkultur*, p. 153.
16. Kupffer, *Gustav Wyneken*, pp. 223–9.
17. Brodersen, *Walter Benjamin*, pp. 26–7.
18. KdU §§ 1–5.
19. Wyneken, *Jugendkultur*, p. 11.
20. Cf. Brodersen, *Walter Benjamin*, p. 27.
21. *Jugendkultur*, p. 10.
22. For the interrelatedness of 'autonomy' and 'leadership' in Wyneken's thought, see Kupffer, *Gustav Wyneken*, pp. 305–13; and Hillach, 'Lebensgesetz', pp. 875, 884.
23. Cf. Kupffer, *Gustav Wyneken*, pp. 187–93.
24. *Jugendkultur*, p. 10.
25. Letters to Wyneken, 9 March 1915, and Hans Reichenbach, February–March 1915, GB I, 262–4. See also Kurt Flasch, *Die geistige Mobilmachung*, pp. 344–52.
26. Friedrich Nietzsche, *Die Geburt der Tragödie aus dem Geiste der Musik*, pp. 40, 131–4 (trans. pp. 49–50, 183–7); Pfotenhauer, 'Benjamin und Nietzsche', pp. 100–6.
27. See Letter to Strauß, 7–9 January 1913, GB I, 83; and Deuber-Mankowsky, *Benjamin und Cohen*, pp. 297–8, 336–40.
28. Wyneken, *Jugendkultur*, p. 34.
29. For the importance of Benjamin's understanding of early Christianity for his notions of *Literat* and 'youth', see GB I, 64 and GS II.1, 74.
30. Fritz K. Ringer, *The Decline of the German Mandarins*, pp. 253–304; McCole, *Antinomies*, pp. 40–5.
31. See section 3.4.1.
32. Cf. *Jugendkultur*, pp. 159–62.
33. Letter to Carla Seligson, 30 April 1913, GB I, 92.
34. GS II.1, 30. Benjamin's phrase 'Kultur des Ausdrucks' ('culture of expression') represents a semantically neutral modification of the term 'Ausdruckskultur' in the subtitle of the journal *Der Kunstwart und Kulturwart: Halbmonatsschau für Ausdruckskultur auf allen Lebensgebieten*. Both Wyneken and Goldstein published in this influential journal.

35. Cf. *Jugendkultur*, pp. 153, 160–2.
36. Kupffer, *Gustav Wyneken*, pp. 36–9.
37. See also Letters to Herbert Blumenthal, 23 June 1913, Carla Seligson, 15 September 1913; and Ernst Schoen, 23 May 1914. GB I, 124, 175, 230–1.
38. Friedrich Schiller, *Über die ästhetische Erziehung des Menschen in einer Reihe von Briefen*, esp. letters 2–4 and 27.
39. GS II.1, 147, SW I, 67. Cf. Wyneken, *Jugendkultur*, p. 153.
40. See, also for what follows, *Jugendkultur*, pp. 153–5, 161.
41. *Unendliche Verdopplung*, p. 30.
42. 36, incl. note 60, 39–40; *132, 189, note 63, 134–5.*
43. For more balanced discussions, which acknowledge the importance of the structure of the beautiful *object* in Kant's *Third Critique*, see Theodore Edward Uehling, *The Notion of Form in Kant's 'Critique of Aesthetic Judgment'*; and Jens Kulenkampff, *Kants Logik des ästhetischen Urteils*, pp. 112–35.
44. 71–3, 87; *154–6, 165.*
45. For a historical explanation of Schlegel's affirmation of the paradoxical, see my article 'Vereinigung und Paradox'.
46. 63; *150*. Cf. KA II, 204–5 and section 2.2.3. Apart from *Athenäum* fragment 116, Menninghaus (*Verdopplung*, pp. 61, 258, incl. note 46) counts only two, posthumously published, occurrences of 'poetic reflection' in Schlegel. The expression figures in a series of notes, dating from 1806, in which Schlegel subordinates criticism and poetic thought to what he calls 'the categorical decision of moral or divine reflection, that is, of conscience' (KA XIX, 164). The moral dimension, which characterises this decisionistic and religious view, is also present in the earlier note, dating probably from 1797, which has 'poetic reflection'. Schlegel there suggests that the essence of sentimentality consists in poetic reflection upon a moral dualism (KA XVI, 256).
47. KA XII, 107–480 and KA XIII, 3–175. Benjamin relies in particular on books II–V, which Schlegel placed under the heading 'Psychology as a Theory of Consciousness'.
48. See section 2.2.4. I follow established conventions in translating Fichte's 'Ich' by 'I' but choose 'ego-less', rather than 'I-free', 'I-less' or 'without the I', to translate Benjamin's 'Ich-frei'. Hopefully, this sacrifice of terminological continuity to considerations of style will not lead to confusion.
49. GS II.1, 106, SW I, 19. See Caygill, *Colour*, pp. 36–40 and my article 'Heinles Verklärung'.
50. 40; *135*. For Benjamin's notion of criticism in relation to the post-Kantian problem of system see Philippe Lacoue-Labarthe, Jean-Luc Nancy, *The Literary Absolute*, pp. 27–37, 46–9; and Deuber-Mankowsky, *Benjamin und Cohen*, pp. 112–20.
51. Benjamin seems to consider Early Romanticism a (potential) system in the sense outlined above (48–9; *140*). See also section 2.3.2.
52. 11–12, 39; *116–17, 133–4*. See Pierfrancesco Fiorato, '*Problemgeschichte* messianica'; and, for a helpful introduction to the history and problems of *Problemgeschichte*, L. Geldsetzer, W. Hübener, S. Haubold, 'Problemgeschichte', HWP VII.
53. As in German Idealism, in Benjamin 'bestimmen' is frequently used as a German equivalent of the Latin 'determinare' in early modern philosophy.

In most cases, the English translation by 'determine' seems preferable to the more 'readable' but inappropriate alternatives 'define' and 'characterise'.

54. The translation 'must be designated and summarised' is also correct.
55. Benjamin tends to consider Schlegel's interest in theoretical philosophy as one in philosophy *tout court* (22, 44; *123, 137*). For the moral dimension of Schlegel's thought, see Klaus Briegleb, *Ästhetische Sittlichkeit*, pp. 43–156, 223; Franz Norbert Mennemeier, *Friedrich Schlegels Poesiebegriff*, pp. 17–108 and note 46, p. 197 above.
56. 36, 47; *132, 140*. For Benjamin's use of 'zitieren', see GS V.1, 595 (sheaf 'N' from the Arcades project) and GS I.2, 694 ('On the Concept of History').
57. See *Athenäum* fragment 242 (KA II, 205) and section 2.3.2.
58. *fragments*, p. xxi.
59. *Verdopplung*, p. 57.
60. Detlev Schöttker, *Konstruktiver Fragmentarismus*, pp. 12, 55; Philippe Lacoue-Labarthe, 'Introduction to Walter Benjamin's "The Concept of Art Criticism in the German Romantics"', pp. 422, 427; David Ferris, '"Truth is the death of intention"', p. 455.
61. *The Idea of a Critical Theory*, pp. 94–5. Geuss rests his hope on what he calls Adorno's historicism.
62. The translation 'Fichte strives for' is also correct.
63. 'Begriff', p. 49.
64. Cf. 'Begriff', pp. 49, 51, 67, 70.
65. 'Begriff', p. 74.
66. For Fichte's concern with the unity of theoretical and practical reason, see Frederick Neuhouser, *Fichte's Theory of Subjectivity*.
67. 'Begriff', pp. 67, 72.
68. See also Benjamin's quotation, in a different context, from Fichte's *Grundlage der gesammten Wissenschaftslehre* (1794), in which he omits precisely that clause according to which reflection results from the intervention of reason – indeed, from its 'ins *Mittel* Treten' (my emphasis) – in the 'Unmittelbarkeit' ('immediacy') of imagination (22; *123–4*). Cf. *Grundlage*, pp. 216–17 (trans. p. 194).
69. Letter to Fritz Radt, 14 May 1915, GB I, 266; Cassirer, *Substanzbegriff und Funktionsbegriff*, p. 428, trans. (modified) p. 322.
70. See Cohen, *Kants Theorie der Erfahrung*, pp. 245–310, 757–8, 770; Fiorato, 'Die Erfahrung, das Unbedingte und die Religion'; Deuber-Mankowsky, *Benjamin und Cohen*, pp. 55–89; and Natorp, 'Kant und die Marburger Schule', p. 201.
71. *Wahrheit*, pp. 27–34.
72. Cf. also for what follows, Menninghaus, *Verdopplung*, p. 34.
73. *Grundlage*, pp. 91–2 (trans. p. 93).
74. *Grundlage*, p. 295 (trans. p. 259).
75. *Funktionsbegriff*, p. 430 (trans. p. 323).
76. Menninghaus, *Verdopplung*, pp. 31–2.
77. 25; *125*. Cf. Fichte, 'Versuch einer neuen Darstellung der Wissenschaftslehre', pp. 526–7.
78. *Funktionsbegriff*, p. 450 (trans. p. 338).
79. *Die Geschichte der neueren Philosophie in ihrem Zusammenhange mit der*

allgemeinen Kultur und den besonderen Wissenschaften, II, p. 211; 'Urprinzip' is not emphasised in the original.

80. Heidegger (*Sein und Zeit*, p. 88, trans. p. 122) will go further and criticise theories of functional conceptualisation as blind to their own metaphysical nature; such theories, according to him, work with formalised concepts of substance.
81. Natorp (*Einleitung in die Psychologie nach kritischer Methode*, pp. 11–14) discusses the metaphorical character of 'centre' when used with reference to the I. His use of 'relation' and its cognates strongly resembles Benjamin's.
82. Meister Eckhart, 'Prologus generalis in opus tripartitum', 17, Eckhart, *Expositio libri Genesis*, 7–8; Luther, 'Sermon und Eingang in das erste Buch Mose', p. 441, Luther, *Auslegung des ersten und zweiten Kapitels Johannis*, pp. 558–61; Rosenzweig, *Der Stern der Erlösung*, p. 132 (trans. p. 111) and passim.
83. *Funktionsbegriff*, p. 20 (trans. pp. 15–16). With his notion 'generating relation', Cassirer may be responding to Cohen's programmatic claim (*Kants Theorie der Erfahrung*, pp. 757–8) that the drivel ('das Fabeln') about the spontaneous, auto-creative production of the real has to be translated into the scientific language of the methodological generation of the real.
84. For what follows, see 'Begriff', pp. 48, 74–6 (incl. note), 79–80, 52–3.
85. *Die Theorie des Romans*, pp. 21–30, 107 (trans. pp. 29–39, 130); Michael Makropoulos, 'Haltlose Souveränität', pp. 201, 206–7.
86. 32–3; *130*. Cf. KA XII, 355–6.
87. *Grundlage*, pp. 216–17, 225, 232–3 (trans. pp. 194, 200–1, 206–7).
88. KA XII, 23–6; and *Athenäum* fragments 76, 342, 443, 448 (KA II, 176, 226, 254). See also Eberhard Huge, *Poesie und Reflektion in der Ästhetik des frühen Friedrich Schlegel*, pp. 123–4 and Menninghaus, *Verdopplung*, pp. 43–5.
89. Cf. Novalis ['Fichte-Studien'], pp. 113–25, 206–13; Schelling, *Ideen zu einer Philosphie der Natur*, pp. 13–14; and Fichte, 'Zweite Einleitung in die Wissenschaftslehre', esp. p. 459 (trans. pp. 34–5). For Hölderlin's related view of reflection as necessarily dividing the unity of subject and object given in intellectual intuition, see Andrew Bowie, *Aesthetics and Subjectivity*, pp. 67–72.
90. For a different view, see Menninghaus, *Verdopplung*, p. 36.
91. GS I.1, 195, 178, SW I, 351, 338 and GS I.1, 217.
92. *Antinomies*, p. 32.
93. Lacoue-Labarthe, 'Introduction', p. 422. Cf. Letter to Schoen, 7 April 1919, GB II, 23.
94. 24–5; *124–5*. Cf. Fichte, 'Versuch', pp. 525–6.
95. 'Die Wissenschaftslehre' (1812), pp. 325–6.
96. 'Bericht über den Begriff der Wissenschaftslehre und die bisherigen Schicksale derselben', pp. 364–5.
97. For the problematic translation of Hölderlin's 'genau' by 'precisely' see note 104, p. 200 below.
98. See also '*Trauerspiel* and Tragedy' (1916) and the discussions in Schwarz Wentzer, *Bewahrung*, pp. 74–92; and Caygill, 'Benjamin, Heidegger and the Destruction of Tradition', pp. 4–12.
99. In my discussion above, I disregard Schwarz Wentzer's view that Benjamin's messianism is itself philosophical, and not religious, in kind. Schwarz Wentzer both distinguishes philosophy from theology and works with a

pre-modern conception of philosophy for which its separation from theology is negligible. See *Bewahrung*, pp. 10, 50, 66, 84–7, 92 and *passim*.

100. 12–3, incl. note 3; *177, 185, note 3*. Letters to Schoen, 8–9 November 1918 (GB I, 486) and 7 April 1919 (GB II, 23).
101. 12, 91–3; *117, 167–8*.
102. *Grundlage*, pp. 215–17 (trans. pp. 193–5). For what follows, see also Menninghaus, *Verdopplung*, pp. 53–4.
103. Rosenzweig, *Das älteste Systemprogramm des deutschen Idealismus: Ein handschriftlicher Fund*, pp. 5–7 (trans. in Bowie, *Aesthetics and Subjectivity*, pp. 265–7). See also Nancy Lacoue-Labarthe, *The Literary Absolute*, pp. 27–37 and, for the question of authorship, Rüdiger Bubner (ed.), *Das älteste Systemprogramm*.
104. The translation of Hölderlin's 'genau' by 'precisely' is misleading if Hölderlin, as is likely, uses 'genau' in the old sense of 'nah' ('closely'). (Cf. 'genau', DW V.) However, a translation of 'genau' by 'closely' would make Benjamin's interpretation of 'genau' as meaning 'systematically' look even more problematic.
105. Hölderlin, [*Neun Pindar-Kommentare*], pp. 277–8, 294–5; and Letter to Scholem, 11 November 1916, GB I, 344–6.
106. For the translation of Pindar's phrase, in the *Stephaniana* of 1560, by 'das meine so verschanzend das-Leben-hinbringe' ('thus entrenching what is mine, I-spend-my-life'), see [*Neun Pindar-Kommentare*], p. 294. For the importance of 'Zusammenhang' in Hölderlin's understanding of religion, see Hölderlin, 'Über Religion', esp. p. 275.
107. 27; *126*. Cf. KA II, 139 (Goethe review).
108. See Benjamin's note 60 (*note 63*) and the announcement of his thesis, GS I.2, 707.
109. For Benjamin's use of 'absolute' as an attribute of *Zusammenhang*, see 'On Perception in Itself' (appr. 1917), GS VI, 32.
110. 26, 45; *126, 138*. Cf. Schlegel's letter to his brother August Wilhelm of 28 August 1793: 'What we call *soul* in works, actions, and works of art [...] spirit and moral dignity in man, God in Creation, namely a *Zusammenhang* utterly endowed with life, is, in concepts, the system' (KA XXIII, 129–30). Benjamin's quotes another passage of this letter in his discussion of Schlegel's mysticism (40; *135*).
111. GS II.1, 203–4. See the editors' comments in GS II.3, 946–8, Irving Wohlfarth, '"Immer radikal, niemals konsequent ..."'; and section 2.3.2.
112. 27–30, 54–8, incl. note 141; *127–9, 144–6, 191–2, note 146*.
113. 27; *127*. Benjamin pleads 'terminological' reasons for considering first-order thought self-reflection. Since in any discursive argument, we need substantive, and not terminological, reasons for considering something an instance of a particular notion, he may be using 'terminological' in the sense of 'mystical'. See 47–8, 27–8, incl. note 35; *139–40, 127, 187, note 27*. Cf. Novalis, 'Logologische Fragmente', pp. 525–6.
114. 31–2; *130*. Cf. KA XII, 349 (Cologne lectures).
115. See Karl Heinz Volkmann-Schluck, 'Novalis' magischer Idealismus'; Menke, *Sprachfiguren*, pp. 406–7; Menninghaus, *Verdopplung*, pp. 59–60; and, for the importance of 'magical idealism' in Benjamin's historical thought and 'anthroplogical materialism', 'On the Philosophy of History of Late

Romanticism and the Historical School' (appr. 1921) and 'On Experience' (1931–32). GS VI, 95–7 and 88–9, SW I, 184–5, SW II, 553.

116. 23; *124*.
117. Fichte, 'Versuch', p. 530.
118. 'Versuch', pp. 529–30.
119. *Logische Untersuchungen*, II, pp. 354–63, esp. the notes (trans. pp. 541–51), *Ideen zu einer reinen Phänomenologie und phänomenologischen Philosophie*, I, pp. 194–6 (trans. pp. 172–3). See also David Bell, *Husserl*, pp. 163–8.
120. Menninghaus (*Verdopplung*, pp. 47–9, 115) considers Benjamin's claim of the priority of reflection an anticipation of Jacques Derrida's notion of *différance* and an attempt to deny the 'metaphysics of presence' in the language of that metaphysics. Contrary to his suggestion, however, the notion of counter-sense, in this context, does not suffice to rule out nonsense. Nietzsche's 'God is dead' may well be both an intended denial of religion in the spiritually gratifying language of religion and a nonsensical statement. For a masterly, post-traditional (non-)prayer for apocalypse, see Derrida, 'Of an Apocalyptic Tone Newly Adopted in Philosophy', for an assessment of Derrida's writing as different from negative theology, Menke, *Sprachfiguren*, pp. 17–18.
121. For more fun with dogs and the infinite, see Thomas Bernhard, *Die Billigesser*.
122. See GB I, 63–4 (Letter to Strauß, 11 September 1912) and GS III, 559 (review of Albert Béguin, *L'Âme romantique et le rêve*, 1939).
123. 73, 84, 87; *156, 163, 165*.
124. Menninghaus, *Verdopplung*, p. 33.
125. 84, 87; *163, 165*. See also Benjamin's source Max Pulver, *Romantische Ironie und Romantische Komödie*, esp. pp. 1–9; and Ernst Behler, 'Die Theorie der romantischen Ironie im Lichte der handschriftlichen Fragmente Friedrich Schlegels'; Ingrid Strohschneider-Kohrs, *Die romantische Ironie in Theorie und Gestaltung*, pp. 14–54; Karl Konrad Polheim, *Die Arabeske*, pp. 218–22; Mennemeier, *Schlegels Poesiebegriff*, pp. 92, 233, 249–51; Menninghaus, *Verdopplung*, pp. 196–207; Manfred Frank, *Einführung in die frühromantische Ästhetik*, pp. 307–16; and Marike Finlay, *The Romantic Irony of Semiotics*.
126. KA XVIII, 476 and Menninghaus, *Verdopplung*, pp. 196–203.
127. Cf. Friedrich Gundolf, *Romantiker*, p. 45; and, more recently, Behler, 'Nietzsche und die frühromantische Schule'.
128. *Verdopplung*, p. 203.
129. *Ion*, 533d-534b.
130. See KA XVI, 6 ('Von der Schönheit in der Dichtkunst'), KA I, 403–5 (*Geschichte der Poesie der Griechen und Römer*) and that passage of the so-called 'Studium-Aufsatz' – KA I, 274–5 – from which Benjamin himself quotes (114; *181–2*).
131. Jochen Schmidt, *Die Geschichte des Geniegedankens*, I, pp. 132–5.
132. For the importance of 'classical' in Schlegel's poetological thought, see Annelen Grosse-Brockhoff, *Das Konzept des Klassischen bei Friedrich und August Wilhelm Schlegel*.
133. *Phaidros*, 249c8–d3.
134. 86; *165*. Cf. Schleiermacher, *Der christliche Glaube nach den Grundsätzen der*

evangelischen Kirche im Zusammenhange dargestellt, pp. 31–187, and GS II.1, 33 ('Religiosity of the Present').

135. Lieselotte Wiesenthal, *Zur Wissenschaftstheorie Walter Benjamins*, pp. 10–17; Heil, '*Gefährliche Beziehungen*', pp. 156–61.
136. For discussions which adopt Benjamin's interpretation of 'immanent criticism' in their analyses of his critical practice, see, for instance, Caygill, *Colour*, pp. ix, 35–52; and McCole, Antinomies, pp. 71–114.
137. By 'Darstellung', which frequently means 'representation', Benjamin may refer to a (quasi-)chemical 'preparation' ('Darstellung'). See section 2.3.2.
138. Cf. KA II, 349 (*Gespräch über Poesie*) and KA XIX, 164 (*Philosophische Lehrjahre*).
139. 14; *118*. Cf. the assessment of Schlegel's unfinished piece as the best review ever written in the German language in Gundolf, *Shakespeare und der deutsche Geist*, pp. 328–9.
140. *Verdopplung*, pp. 186–95.
141. See *Lyceum* fragment 112 (KA II, 161), Willy Michel, *Ästhetische Hermeneutik und frühromantische Kritik* and GS IV.1, 9, SW I, 253 ('The Task of the Translator').
142. Schlegel's interpretation of Goethe's characters as ciphers of the novel for the novel itself (KA II, 130, 145, 143) – arguably the most important single device by which the review evokes the idea that *Apprenticeship* reflects itself – ascribes to Goethe that allegorical technique which Schlegel himself employs in his unfinished novel *Lucinde*. With regard to a theory of literary self-reflection, this interpretation is problematic, for the referents of Schlegel's allegorical *quid pro quo* – the work, the genius of the work, the world spirit of *Apprenticeship* – do not figure in any established allegorical convention on which Goethe's novel would demonstrably rely. See KA V, 19–27 (*Lucinde*), Letter to Caroline Schlegel, March 1799 (KA XXIV, 252–3); and the analyses in Michel, *Ästhetische Hermeneutik*, pp. 261–90; Menninghaus, *Verdopplung*, pp. 174–83, 186–95; and Mennemeier, *Schlegels Poesiebegriff*, pp. 221–63.
143. KA II, 131. The translation of 'durchaus organisiert' by 'somewhat organised' may also be correct. Cf. Schlegel's only posthumously published assessment 'Goethe had no idea of romantic wholeness' (KA XVI, 133) and the third point made above.
144. Cf. McCole, *Antinomies*, pp. 147–55; Steiner, '*Zarte Empirie*'; and sections 3.2.1 and 3.4.1.
145. For Benjamin's simultaneous affirmation and implicit negation of the immanence of the idea in 'The Life of Students' and the Hölderlin essay, see my discussions in 'The Desire for Objectivity', pp. 34–6; and 'Heinles Verklärung', pp. 141–2.
146. In 'The Task of the Translator', Benjamin considers 'pure language' the object of *Darstellung* (GS IV.1, pp. 11–13, SW I, 255–6). For the importance of '*Darstellung* of the prosaic core in every work' in 'On the Concept of History' see section 3.5.
147. See *Athenäum* fragments 83, 220, 404, 412, 426, 439 (KA II, 178, 200, 241, 243, 248, 253).
148. *Athenäum* fragment 242 starts with the suggestion that ancient literature ('die alte Poesie') is an individual (KA II, 205), and the rejection, in fragment

83, of the principle of non-contradiction would be futile if 'individual' did not imply a reference to language.

149. For the importance which Benjamin attributed to the formal composition ('Komposition') of *The Concept of Art Criticism*, see Letter to Schoen, 14 May 1919, GB II, 26.
150. For Benjamin's reliance on Hölderlin's coinage 'heilignüchtern' ('holy-sober'), see 103–6 (*175–6*). In contrast to Benjamin's image of a destructive, sober and, implicitly, holy light, however, Hölderlin's 'heilignüchtern' lacks the connotation of destruction. Where Hölderlin is concerned with the danger which man faces in encountering the sacred, he uses an imagery of light but not 'heilignüchtern'. See *Gedichte nach 1800*, pp. 117, 118–20, 90–5, 165–87 (trans. *Poems and Fragments*, pp. 370–7, 242–53, 462–87).
151. 'Shakespeare und unsere Religion', p. 6, quoted in Steiner, *Geburt der Kritik*, p. 190. Wolin (*Walter Benjamin*, xxix–lii) interprets Benjamin's 'nihilism' as part of a 'secular messianism', Franco Moretti (*Signs Taken for Wonder*, pp. 240–61) places it against the background of 'Left' cults of revolutionary apocalypse.

Chapter 3

1. For the problem of continuity in the history of Benjamin reception, see McCole, *Antinomies*, pp. 10–21.
2. See, for instance, Axel Honneth, 'A Communicative Disclosure of the Past', pp. 93–4; H. D. Kittsteiner, 'Walter Benjamin's Historicism', pp. 207–15; Hannah Arendt, 'Introduction', pp. 17–19; Esther Leslie, *Walter Benjamin: Overpowering Conformism*, pp. xiii, 150–61, 234; and Weigel, *Body- and Image-Space*, pp. 3–60, esp. 10, 18.
3. For non-religious accounts of the performative dimension of Benjamin's (later) writing, see Schöttker, *Konstruktiver Fragmentarismus*; Heiner Weidmann, 'Geistesgegenwart'; Eva Geulen, 'Zeit zur Darstellung'; Burkhardt Lindner, 'Benjamins Aurakonzeption', pp. 239–44; and McCole, *Antinomies*, pp. 30–4, 156–80. With recourse to Paul de Man's notion of performance, Menke (*Sprachfiguren*, esp. pp. 18, 395–404) construes Benjamin's writing as self-referential, performative *Darstellung*. For a performative comparison between Benjamin and Jean-François Lyotard, see Kenneth Lea, 'Traducing History', pp. 127–37.
4. Cf. Letter to Rang, 9 December 1923, GB II, 393, SW I, 389.
5. GS I.1, 211. Cf. GS I.1, 126, SW I, 298 ('Goethe's Elective Affinities').
6. See, also for what follows, GS I.1, 89–91, SW I, 166–7 (*The Concept of Art Criticism*); and Schwarz Wentzer, *Bewahrung*, pp. 215–19, 232.
7. See also Steiner, *Geburt der Kritik*, p. 271.
8. *Geburt der Kritik*, p. 271.
9. GS II.1, 140–1, SW I, 62 ('On Language'). Benjamin implicitly relates the absolute to history when claiming that there is no event ('Geschehen') that does not partake of language.
10. Bowie (*From Romanticism to Critical Theory*, p. 213) argues that in the thesis Benjamin's central concern is already 'the real historical process of reflection, in which certain works do indeed continue to provoke further

reflection'. The phrase 'the real historical process of reflection' can mislead about both the striking formalism of Benjamin's thesis and the fact that his later concern with the post-history of the works may not be empirical in kind.

11. The growing historical work on the Frankfurt School and its context suggests that the widespread assumption that an author's political or existential stance is closely linked to the theories which he endorses needs to be reassessed. See, for instance, Caygill, 'Benjamin, Heidegger and the Destruction of Tradition'; Heil, '*Gefährliche Beziehungen*'; Horst Bredekamp, 'From Walter Benjamin to Carl Schmitt, via Thomas Hobbes'; Bowie, 'The Romantic Connection'; and Manfred Gangl, Gérard Raulet (eds), *Intellektuellendiskurse in der Weimarer Republik*.
12. See GS I.1, 360, 363, 387–9; McCole, *Antinomies*, p. 133, note 19; and GS III, 557–60 (Béguin review).
13. See also Stoessel, *Aura*, pp. 56–7; and, for different views, Françoise Meltzer, 'Acedia and Melancholia', pp. 161–2; and Lindner, 'Engel und Zwerg', pp. 242–5, 259–60. For other early twentieth-century versions of 'manly romanticism', see Hugo von Hofmannsthal, *Das Schrifttum als geistiger Raum der Nation*, pp. 37–9; the extract of Rang's letter to Benjamin quoted in GB II, 396–7; and Gisela Dischner, 'Freiheit auf dem Weg zur Entsagung? Zu Otto Weiningers *Geschlecht und Charakter*'.
14. Hans-Jürgen Schings, 'Consolatio Tragoediae', p. 22; Schings, 'Walter Benjamin, das barocke Trauerspiel und die Barockforschung'; Kittsteiner, 'The Allegory of the Philosophy of History in the Nineteenth Century', pp. 55–6; Klaus Garber, *Rezeption und Rettung*, pp. 59–120, Garber, 'Benjamin und das Barock'; Witte, *Der Intellektuelle*, pp. 111–12; Steiner, 'Allegorie und Allergie'.
15. Mieth, 'Fremdheit Gottes', pp. 108–10.
16. Letter to Scholem, 22 December 1924, GB II, 508–13.
17. For what follows see also Brodersen, *Walter Benjamin*, pp. 135–91.
18. Cf. Letter to Max Rychner, 7 March 1931, GB IV, 17–20.
19. McCole, *Antinomies*, pp. 156–205. McCole (p. 197) disagrees with my understanding of Benjamin's Marxism as atheoretical.
20. As a strategy, and not a theory or direct political conviction, Benjamin's 'Marxism' conforms to both his earlier advocacy of an alliance between the *Jugendkulturbewegung* and the Left and the view, in his notes on surrealism of 1928–29, that political writings express convictions less often than one is inclined to think. GB I, 83 (Letter to Strauß, 7–9 January 1913) and GS II.3, 1033.
21. Letter to Scholem, 29 May 1926, GB III, 159.
22. GB IV, 27–8. Cf. Arendt, 'Introduction', pp. 41–2.
23. Letter to Scholem, 29 May 1926, GB III, 158–63. In this letter, Benjamin also denies the existence of 'meaningfully *political* objectives' and states his belief in the paradoxical identity of political and religious commitment.
24. See also William James, 'The Will to Believe'.
25. See also for what follows, GS II.1, 297–310, SW II, 208–18 and GS II.3, 1039–41.
26. Korsch, *Marxismus und Philosophie* and, for a different assessment, Leslie, *Overpowering Conformism*, p. 97.
27. For a different assessment, see McCole, *Antinomies*, p. 227.

28. GS IV.1, 146–8; SW I, 486–7 (*One-Way Street*); and GS III, 247–8, SW II, 318–19 ('Theories of German Fascism', 1930). For accounts which emphasise the contrast between Benjamin's and conservative visions of *Technik*, see McCole, *Antinomies*, pp. 179–90, Leslie, *Overpowering Conformism*, pp. 1–41, and Pivecka, *Künstliche Natur*, pp. 32–45.
29. See GB II, 127–8 (Letter to Scholem, January 1921); GS II.1, 190–1, SW I, 243–4 ('Critique of Violence'); Manfred Voigts, 'Walter Benjamin und Erich Unger', Lindner, 'Derrida, Benjamin, Holocaust', p. 1703, note 10, and sections 3.4.2–3.
30. GS VI, 64–7, 78–87; SW I, 393–400.
31. GS V.2, 800–23; and, for a different assessment, Leslie, *Overpowering Conformism*, p. 40. In the Marx file, Benjamin focuses on those aspects of Marx's early writings and *Capital* which are compatible with a romantic conception of man's relation to nature and the privileging, in German Idealism, of activity over substance. See esp. Marx, 'Zur Kritik der Nationalökonomie', pp. 566–9, 593–4 (trans. 75–7, 84).
32. Cf. Schlegel's 'Ideas', 24, 64, 95, 152 (KA II, 258, 262, 265, 273).
33. GS II.1, 309–10; SW II, 217–18. Cf. GS I.3, 1239. *Aktualität*, which in common German means 'topicality', in Benjamin frequently refers to the full presence of the highest reality in the state of redemption. This usage is historically related to the scholastic notion of God as the only *actus purus*, which, presumably via Fichte, also influenced the use of 'attualità' and 'Aktualität' in the works of Benjamin's contemporaries Giovanni Gentile and Wilhelm Wundt. Cf. D. Schlüter, 'Akt/Potenz', Liesel Königs, 'Aktualismus', and Königs, 'Aktualitätstheorie', HPW I. See also Weigel, *Body- and Image-Space*, pp. 3–15 and McCole, *Antinomies*, pp. 23–5.
34. Cf. McCole, *Antinomies*, p. 180; and Leslie, *Overpowering Conformism*, pp. xi-xiii.
35. GS II.1, 191–2; SW I, 244–5. Where 'Critique of Violence' has 'means' in conjunction with 'pure', 'means' translates the Latin term 'medium' as Benjamin used it in 'On Language' in contradistinction to 'means'. (Cf. Menke, *Sprachfiguren*, pp. 42–5; and Werner Hamacher, 'Afformative, Strike', p. 129, note 13.) The conception of pure means as 'mediatory solutions', or 'technical' arbitrations via things, moreover, corresponds to a note of 1919–20, according to which the divine manifests itself not in the social sphere but in revelatory perception and holy language (GS VI, 99). The technical 'mediation' via things, we may clarify, coincides with the immediate revelation '*in*' things.
36. McCole, *Antinomies*, pp. 180–6; Schöttker, *Konstruktiver Fragmentarismus*, pp. 145–203; Leslie, *Overpowering Conformism*, passim.
37. *Walter Benjamin*, p. 108.
38. GS III, 287–90, SW II, 462–4 ('Literary History and Literary Research'); GS III, 259, SW II, 383 ('Against a Masterpiece', 1930); GS III, 363 ('Someone holds', 1932); GS III, 319–21, SW II 572–4 ('Privileged Thinking', 1932), GS VI, 164–7, 169–79, SW II, 292–5, 402–10 (notes on 'criticism' of 1929–31).
39. Cf. McCole, *Antinomies*, pp. 173–5. Benjamin's opponents in 'Literary History and Literary Research' – like, incidentally, those in 'On the Concept of History' who in their 'brothel of historicism' are not man enough to blast open the *continuum* of history – share the misfortune of 'the romantic

aestheticians' of the *Trauerspiel* study of being both effeminate and lecherous (GS III, 286, SW II, 461, GS I.2, 702). A note from the summer of 1934, in which Benjamin takes issue with Hitler's 'diminished masculinity' points to the political context of Benjamin's polemics (GS VI, 103–4, SW II, 792–3).

40. See Letter to Scholem, 6 May 1934, GB IV, 410, Chryssoula Kambas, *Walter Benjamin im Exil*, pp. 16–80; and the editors' comments in GS II.3, 1460–3.
41. See, also for what follows, GS II.2, 684–6, 701, SW II, 769–70, 780.
42. In 1932, Lukács ('Tendenz oder Parteilichkeit', pp. 18, 20–1) proclaimed Balzac and Tolstoy as the relevant models for socialist literature. His 'Größe und Verfall des Expressionsmus', which anticipated the so-called Expressionism Debate and the eventually brutal communist suppression of modernist and avant-guardist art, was published in January 1934, a few months before the conception of 'The Author'.
43. Peter Bürger, *Theorie der Avantgarde*, pp. 38–40 (trans. pp. 27–30); McCole, *Antinomies*, p. 194; Habermas, *Technik und Wissenschaft als 'Ideologie'*, pp. 54–8 (trans. pp. 86–91).
44. See also GS VI, 178–9, SW II, 407–8 ('False Critique').
45. See, also for what follows, GS IV.1, 147, SW I, 486–7 (*One-Way Street*), GS III, 247–8, SW II, 318–20 ('German Fascism'), GS I.2, 444–5, 468–9, 507–8, GS VII.1, 359–60 ('The Work of Art').
46. GS II.1, 309–10, SW II, 217–18 ('Surrealism'); McCole, *Antinomies*, p. 194.
47. Eagleton, *The Ideology of the Aesthetic*, pp. 336–7; Bürger, *Avantgarde*, p. 38 (trans. p. 30); Rochlitz, *Disenchantment*, p. 176.
48. For the immediacy of Benjamin's 'materialism', see the widely discussed controversy between Adorno and Benjamin in their correspondence of November and December 1938 (AB, 364–86). As I perceive it, this controversy does not concern issues peculiar to dialectics, materialist or not, but the difference between mysticism and *any* discursive theory (with dialectics perhaps occupying a middle ground). To some extent, the two exiled writers in 1938 re-enact one of the most 'auratic' events of traditional German culture: the 1794 debate between Schiller and Goethe concerning the epistemic status of *Urphänomene*.
49. For Benjamin's discussion of 'New Sobriety' and 'Activism', see Kambas, *Im Exil*, pp. 46–63 and section 3.4.2.
50. GS II.2, 686–7, SW II, 770–1; Kambas, *Im Exil*, pp. 39–45.
51. Cf. the editors' comments in GS II.3, 1464.
52. Novalis, 'Vermischte Bemerkungen', p. 470.
53. Benjamin's 'Moscow Diary' demonstrates his awareness of the culturally conservative 'tendencies' in the Soviet Union in the second half of the 1920s. GS VI, 293–7, 338–9.
54. GS VI, 330–1. For Benjamin's changed use of terms such as 'expressive' and 'the communicative' in texts written after 'On Language', see Menke, *Sprachfiguren*, p. 34. The context of the passage quoted above suggests that Benjamin considered the substance of his thought unchanged.
55. *Arbeitsjournal*, I, p. 16 (25 July 1938): 'mysticism throughout, combined with an antimystical stance. that is how the materialist conception of history is being adapted! it's rather dreadful.'
56. See, also for what follows, GS II.2, 662–3, SW II, 366–7. By characterising

Keuner as someone who is not a 'Kraftmensch' ('muscleman'), Benjamin distinguishes his understanding of 'Führer' from the ideals of both National Socialism and Storm and Stress. The phrase 'he who belongs to all' recalls Hölderlin's characterisation of *Geist* ('Vater Aether') in the elegy 'Brod und Wein'. For Brecht's notion that gestures are 'quotable', see Brecht, *Versuche 1–3*, p. 7; Brecht, [Der Fatzerkommentar], pp. 515–16; John Willett (ed.), *Brecht on Theatre*, p. 42 (note); and Roswitha Mueller, 'Learning for a New Society', pp. 85–90.

57. GS II.2, 689–91, SW II, 772–3 ('The Author'). Cf. GS II.1, 200–2, SW I, 251 ('Critique of Violence') and Benjamin's Hiller review of 1932 (GS III, 350–2). See also McCole, *Antinomies*, p. 174 and, for a corresponding critique of 'Activism', Unger, *Politik und Metaphysik*, pp. 14–22, 26–33, 48.
58. The apparently idealistic focus on a change of attitudes which characterises Benjamin's discussions of Brecht can be explained by the fact that for Benjamin's 'anthropological materialism', as for Unger's spiritual notion of politics, the philosophical alternative between idealism and Marx's historical materialism is itself misleading. (Cf. Unger, *Politik und Metaphysik*, pp. 17–18.)
59. *Poetics*, 1449b24–7, 1450a37–1450b4. See Aryeh Kosman, 'Acting', pp. 51, 57–8.
60. The fact that the German original of the passage under discussion has 'handeln' and not 'Handlung' may lend support to my interpretation.
61. Cf. Leslie, *Overpowering Conformism*, p. 97.
62. GS II.2, 609, SW I, 112 ('Stifter', 1918), GS I.1, 112, SW I, 180 (*The Concept of Art Criticism*), GS I.1, 217 (*Trauerspiel* study). See also 'vernehmen' and 'Vernunft', DW XXV, Arendt, 'Introduction', p. 19; Scholem, *Walter Benjamin*, p. 108 (trans. p. 84), Steiner, *Geburt der Kritik*, pp. 45–50; Stoessel, *Aura*, pp. 81–3; and Rochlitz, *Disenchantment*, pp. 35–6.
63. GS II.2, 699, SW II, 779.
64. See section 3.5. For Benjamin's observation that 'Haltung' ('attitude', literally: 'halting') evokes the notion of interrupion or arrest, see his diary entry of 5 May 1931 (GS VI, 425, SW II, 472) and Rainer Nägele, *Theater, Theory, Speculation*, pp. 135–66.
65. GS VI, 524–6, SW II, 784–5. See also Young-Ok Kim, *Selbstporträt im Text des Anderen*, pp. 241–335; Moses, 'Brecht und Benjamin als Kafka-Interpreten'; Pfotenhauer, Ästhetische Erfahrung, pp. 103–19; and Hans Mayer, 'Walter Benjamin und Franz Kafka'.
66. Brecht, 'Über gestische Musik' (1937–38), pp. 329–30 (trans. p. 104); and Peter Brooker, 'Key Words in Brecht's Theory and Practice of Theatre', pp. 195–9.
67. GS II.2, 418, SW II, 801. Cf. Kafka, 'Ein Brudermord', pp. 163–4 (trans. pp. 190–1).
68. Kafka, 'Konvolut 1920', p. 354 (trans. p. 343).
69. For a good, non-religious account of Epic Theatre as a catalyst for Benjamin's notion of reception in 'The Work of Art', see McCole, *Antinomies*, pp. 194–205. See also Gertrud Koch, 'Cosmos in Film'; Miriam Hansen, 'Benjamin and Cinema'; Leslie, *Overpowering Conformism*, pp. 130–67; and Pivecka, *Künstliche Natur*, pp. 112–30.
70. See note 62 above.

71. Hegel, *Vorlesungen über die Ästhetik*, II, pp. 303–7 (trans. pp. 660–3).
72. For Alois Riegl's influence on Benjamin's notion of the tactile, see Ulrich Rüffer, 'Taktilität und Nähe'; and Schöttker, *Konstruktiver Fragmentarismus*, pp. 244–7. Riegl (*Spätrömische Kunstindustrie*, esp. pp. 9, 31–6) uses 'tactile' not primarily in a theory of reception but in the context of his anti-materialist and 'teleological' conception of artistic production ('Kunstwollen').
73. GS II.3, 1041. For 'dispersion' as a possible connotation of 'Zerstreuung' ('distraction', 'dispersion') in 'The Work of Art', cf. the opposition between Epic Theatre and 'distraction', GS II.2, 697, SW II, 778 ('The Author'), and GS I.1, 212–14 (*Trauerspiel* study). See also Hansen, 'Benjamin and Cinema', pp. 210–19; and Lindner, *Aurakonzeption*, pp. 225–36.
74. For what follows, see GS I.2, 450, 466, 487–8, 503–5, incl. note 29.
75. Recki, *Aura und Autonomie*, pp. 66–7; Pfotenhauer, *Ästhetische Erfahrung*, pp. 83–4; Koch, 'Cosmos in Film', p. 211; Hansen, 'Benjamin and Cinema', pp. 185, 204; Rochlitz, *Disenchantment*, pp.161–2; Pivecka, *Künstliche Natur*, pp. 124–8.
76. At least until the turn of the twentieth century, 'Geistesgegenwart' ('presence of mind') seems to have been used as the opposite of 'Zerstreutheit' ('distractedness', 'absent-mindedness'). See 'Geistesgegenwart', DW V.
77. Cf. Stoessel, *Aura*, pp. 157–62.
78. GS VI, 75–6. I owe this reference to Cathriona Bourke.
79. GS V.1, 598 (sheaf 'N' from the Arcades project) and, also for what follows, Benjamin's notes for 'On the Concept of History', esp. GS I.3, 1231–50.
80. *Athenäum* fragments 80, 249 (KA II, 176, 207).
81. *Geistesgegenwart*, pp. 523–34. See also Kambas, '*Politische Aktualität*'.
82. *Politik und Metaphysik*, p. 19. For what follows, see pp. 3–13, 18–28, 30–3, 42–51.
83. To the extent that Benjamin, in contrast to Unger, emphasises eschatological closure, we may say that Benjamin attempts to construe modern technology as important not to the respiritualisation but to the messianic end of society.
84. GS VII.1, 359 ('The Work of Art').
85. Cf. Bolz, 'Die Schrift des Films', p. 33.
86. 'Cosmos in Film', pp. 211–13.
87. GS II.1, 218, SW II, 734–5 ('Experience and Poverty'). Cf. GS I.2, 462.
88. Cf. Pivecka, *Künstliche Natur*, pp. 26, 125–6.
89. GS I.2, 441–5, 468–9, 480–5, 507–8 and GS VII.1, 357–61.
90. *Minima Moralia*, p. 333, trans. p. 247 (my italics).
91. Letter to Gretel Adorno, April 1940, GS I.3, 1226–7.
92. Cf. 'einhalten', DW III.
93. In Benjamin's French translation, the shock produced in the configuration is called 'une sorte de choc en retour' ('a kind of counter-shock') (GS I.3, 1265). See also Hartmut Engelhardt, 'Der historische Gegenstand als Monade', pp. 308–9.
94. For what follows see 'Abschluß des Lessing-Aufsatzes' (KA II, 410–11), the related passage in *Gespräch über Poesie* (KA II, 349), and section 2.3.1.
95. For a different assessment, see McCole, *Antinomies*, 248–9.

96. For Benjamin's problematic conception of historicism, see Kittsteiner, 'Walter Benjamin's Historicism'; Harro Müller, 'Einige Anmerkungen zu Benjamins Historismuskritik'; and for a relatively recent adoption of that conception, Wohlfarth, 'Smashing the Kaleidoscope'. Benjamin himself uses Leopold von Ranke's notion that each period is directly related to God ('unmittelbar zu Gott') to illustrate the monadological 'explosion' of temporal continuity (GS I.3, 1244).
97. Cf. GS I.1, 100–1, SW I, 173, GS I.3, 1234. See also GS I.3, 1238–9, GS II.2, 451 ('The Storyteller'); and Witte, *'Die Welt allseitiger und integraler Aktualität'*, pp. 36–7.
98. Cf. Pfotenhauer, *Ästhetische Erfahrung*, p. 23.
99. The predicate 'einsteht' in the original can be interpreted as an unusual variation of the 'einhält' in Reflection XVII. Cf. note 92, p. 208 above.
100. GB II, 45. See section 1.3.2.
101. *'Eine Puppe in türkischer Tracht'*, p. 279.

Bibliography

See also References and Abbreviations, p. xvii above.

Adorno, Theodor W., 'Einleitung zu Benjamin's *Schriften*' in Adorno, *Gesammelte Schriften*, vol. 11, ed. Rolf Tiedemann, Frankfurt/M. 1974, pp. 567–82.

Adorno, Theodor W., 'Charakteristik Walter Benjamins' in Adorno, *Prismen: Kulturkritik und Gesellschaft*, ed. Rolf Tiedemann, Frankfurt/M. 1976, pp. 234–49.

Adorno, Theodor W., *Minima Moralia: Reflexionen aus dem beschädigten Leben*, Frankfurt/M. 1976; trans. *Minima Moralia*, E. F. N. Jephcott, London 1974.

Arendt, Hannah, 'Introduction' in Walter Benjamin, *Illuminations*, ed. Hannah Arendt, trans. Harry Zohn, London 1992, pp. 7–55.

Baader, Franz Xaver, von, *Vorlesungen über J. Böhme's Theologumena und Philosopheme* in Baader, *Sämtliche Werke*, vol. 3, ed. Franz Hoffmann, Aalen 1963 (= repr. of the edition: Leipzig 1852), pp. 357–436.

Baillie, John, *The Idea of Revelation in Recent Thought*, New York 1956.

Barkhoff, Jürgen, *Magnetische Fiktionen: Literarisierung des Mesmerismus in der Romantik*, Stuttgart 1995.

Béguin, Albert, *L'Âme romantique et le rêve: Essai sur le romantisme allemand et la poésie française*, 2 vols, Marseilles 1937.

Behler, Ernst, 'Die Theorie der romantischen Ironie im Lichte der handschriftlichen Fragmente Friedrich Schlegels', *Zeitschrift für deutsche Philologie*, 88 (1969), 90–113.

Behler, Ernst, 'Nietzsche und die frühromantische Schule', *Nietzsche-Studien*, 7 (1978), 59–87.

Bell, David, *Husserl*, London 1990.

Bernhard, Thomas, *Die Billigesser*, Frankfurt/M. 1980.

Biale, David, *Gershom Scholem: Kabbalah and Counter-History*, Cambridge (Mass.) 1979.

Blumenberg, Hans, *Säkularisierung und Selbstbehauptung*, Frankfurt/M.1974.

Böhme, Jakob, *De signatura rerum, oder Von der Geburt und Bezeichnung aller Wesen* in *Sämtliche Schriften* (= repr. of the 1730 edition of *Theosophia Revelata, Oder: Alle Göttliche Schriften Jacob Böhmens von Altseidenberg*, ed. Johann Georg Gichtel), vol. 6, ed. Will-Erich Peuckert, Stuttgart 1957, pp. 1–244.

Bolz, Norbert, Willem van Reijen, *Walter Benjamin*, Frankfurt/M. 1991.

Bolz, Norbert, 'Die Schrift des Films' in *Diskursanalysen 1: Medien*, eds Friedrich A. Kittler, Manfred Schneider, Samuel Weber, Opladen 1987, pp. 26–34.

Bowie, Andrew, *Aesthetics and Subjectivity: from Kant to Nietzsche*, Manchester 1993.

Bowie, Andrew, *From Romanticism to Critical Theory: The Philosophy of German Literary Theory*, London 1997.

Bowie, Andrew, 'The Romantic Connection: Neurath, the Frankfurt School, and Heidegger', 2 parts, *British Journal for the History of Philosophy*, 8:2 (2000), 275–98 and 8:3 (2000), 459–83.

Brecht on Theatre: The Development of An Aesthetic, ed. and trans. John Willett, London 1986.

Brecht, Bertolt, *Arbeitsjournal*, vol. 1, ed. Werner Hecht, Frankfurt/M. 1973.

Brecht, Bertolt, *Versuche 1–3* in Brecht, *Versuche 1–12: Heft 1–4*, reprint, Berlin 1959, pp. 5–41.

Brecht, Bertolt, [Der Fatzerkommentar] in Brecht, *Werke*, eds Werner Hecht, Jan Knopf, Werner Mittenzwei, Klaus-Detlef Müller, vol. 10.1, Berlin 1997, pp. 513–29.

Brecht, Bertolt, 'Über gestische Musik' in Brecht *Werke*, ed. Werner Hecht, Jan Knopf, Werner Mittenzwei, Klaus-Detlef Müller, vol. 22.1, Berlin 1993, pp. 329–31; trans. 'On Gestic Music', John Willett, in *Brecht on Theatre*, London 1986, pp. 104–6.

Bredekamp, Horst, 'From Walter Benjamin to Carl Schmitt, via Thomas Hobbes', *Critical Inquiry*, 25:2 (1999), 247–66.

Briegleb, Klaus, *Ästhetische Sittlichkeit: Versuch über Friedrich Schlegels Systementwurf zur Begründung der Dichtungskritik*, Tübingen 1962.

Bröcker, Michael, *Die Grundlosigkeit der Wahrheit: Zum Verhältnis von Sprache, Geschichte und Theologie bei Walter Benjamin*, Würzburg 1993.

Brodersen, Momme, *Walter Benjamin: A Biography*, ed. M. Dervis, trans. Malcolm R. Green, Ingrida Ligers, London 1996.

Brooker, Peter, 'Key Words in Brecht's Theory and Practice of Theatre' in *Brecht*, eds. Peter Thomson, Glendyr Sacks, Cambridge 1994, pp. 185–200.

Bruce, Steve, *Religion in the Modern World: From Cathedrals to Cults*, New York 1996.

Buber, Martin, *Drei Reden über das Judentum*, 2nd edition, Frankfurt/M. 1920.

Buber, Martin, *On Judaism*, ed. Nahum N. Glatzer, New York 1972.

Bubner, Rüdiger (ed.), *Das älteste Systemprogramm: Studien zur Frühgeschichte des Idealismus*, Bonn 1973.

Bubner, Rüdiger, *Ästhetische Erfahrung*, Frankfurt/M. 1989.

Büchner, Georg, *Dantons Tod: Kritische Studienausgabe des Originals mit Quellen, Aufsätzen und Materialien*, ed. Peter von Becker, 2nd edition, Frankfurt/M. 1985.

Bürger, Peter, *Theorie der Avantgarde*, Frankfurt/M. 1974; trans. *Theory of the Avant-Garde*, Michael Shaw, Minneapolis 1984.

Burke, Kenneth, *A Rhetoric of Motives*, Berkeley 1969.

Capelle, Cathérine O. P., 'Introduction' in Saint Thomas Aqunas, *L'Être et L'Essence*, ed. and trans. Capelle, Paris 1965, pp. 5–10.

Cassirer, Ernst, *Substanzbegriff und Funktionsbegriff: Untersuchungen über die Grundlagen der Erkenntniskritik*, Darmstadt 1968 (= repr. of the first edition: Berlin 1910); trans. in *Substance and Function and Einstein's Theory of Relativity: Both Books Bound as One*, William Curtis Swabey and Marie Collins Swabey, New York 1953.

Caygill, Howard, 'Benjamin, Heidegger and the Destruction of Tradition' in *Walter Benjamin's Philosophy: Destruction and Experience*, eds. Andrew Benjamin, Peter Osborne, London 1994, pp. 1–31.

Caygill, Howard, *Walter Benjamin: The Colour of Experience*, London 1998.

Cohen, Hermann, *Kants Theorie der Erfahrung*, 3rd edition, Berlin 1918.

Cohen, Hermann, *Religion der Vernunft aus den Quellen des Judentums: Eine jüdische Religionsgeschichte*, ed. Bruno Strauss, 2nd edition, Wiesbaden 1988.

Cupitt, Don, *After God: The Future of Religion*, London 1998.

Derrida, Jacques, 'Of an Apocalyptic Tone Newly Adopted in Philosophy', trans. John P. Leavey Jr., in *Derrida and Negative Theology*, eds Harold Coward, Toby Foshay, Albany 1992, pp. 25–71.

Deuber-Mankowsky, Astrid, *Der frühe Walter Benjamin und Hermann Cohen: Jüdische Werte, Kritische Philosophie, vergängliche Erfahrung*, Berlin 2000.

Dickie, George, 'The Myth of the Aesthetic Attitude', *American Philosophical Quarterly*, 1 (1964), 56–65.

Dischner, Gisela, 'Freiheit auf dem Weg der Entsagung? Zu Otto Weiningers *Geschlecht und Charakter*' in Weininger, *Geschlecht und Charakter: Eine prinzipielle Untersuchung*, Munich 1980, pp. 656–61.

Dorner, I. A., *System der christlichen Glaubenslehre*, vol. 1: *Grundlegung oder Apologetik*, Berlin 1879.

Eagleton, Terry, *Walter Benjamin or Towards a Revolutionary Criticism*, London 1981.

Eagleton, Terry, *The Ideology of the Aesthetic*, Oxford 1990.

Eckhart (Meister), 'Prologus generalis in opus tripartitum' in Eckhart, *Die lateinischen Werke I*, ed. and trans. Konrad Weiss, Stuttgart 1964, pp. 148–65.

Eckhart (Meister), *Expositio libri Genesis* in Eckhart, *Die lateinischen Werke I*, ed. and trans. Konrad Weiss, Stuttgart 1964, pp. 185–444.

Engelhardt, Hartmut, 'Der historische Gegenstand als Monade: Zu einem Motiv Benjamins' in *Materialien zu Benjamins Thesen 'Über den Begriff der Geschichte'*, ed. Peter Bulthaupt, Frankfurt/M. 1975, pp. 292–317.

Ferris, David, '"Truth is the death of intention": Benjamin's Esoteric History of Romanticism', *Studies in Romanticism*, 31 (1992), 455–80.

Fichte, Johann Gottlieb, 'Über den Begriff der Wissenschaftslehre' in *Fichtes Werke* (= repr. of *Johann Gottlieb Fichtes sämmtliche Werke*, ed. Immanuel Hermann Fichte, 8 vols, Berlin 1845–46 and *Johann Gottlieb Fichtes nachgelassene Werke*, ed. Immanuel Hermann Fichte, 3 vols, Bonn 1834–35), vol. I, Berlin 1971, pp. 27–81.

Fichte, Johann Gottlieb, 'Grundlage der gesammten Wissenschaftslehre' in *Fichtes Werke* (= repr. of *Johann Gottlieb Fichtes sämmtliche Werke*, ed. Immanuel Hermann Fichte, 8 vols, Berlin 1845–46 and *Johann Gottlieb Fichtes nachgelassene Werke*, ed. Immanuel Hermann Fichte, 3 vols, Bonn 1834–35), vol. I, Berlin 1971, pp. 83–328; trans. *Science of Knowledge: With The First and Second Introduction*, eds and trans. Peter Heath, John Lachs, Cambridge 1982.

Fichte, Johann Gottlieb, 'Zweite Einleitung in die Wissenschaftslehre' in *Fichtes Werke* (= repr. of *Johann Gottlieb Fichtes sämmtliche Werke*, ed. Immanuel Hermann Fichte, 8 vols, Berlin 1845–46 and *Johann Gottlieb Fichtes nachgelassene Werke*, ed. Immanuel Hermann Fichte, 3 vols, Bonn 1834–35), vol. I, Berlin 1971, pp. 451–518; trans. 'Second Introduction to the Science of Knowledge' in Fichte, *Science of Knowledge: With The First and Second Introduction*, eds and trans. Peter Heath, John Lachs, Cambridge 1982, pp. 29–85.

Fichte, Johann Gottlieb, 'Versuch einer neuen Darstellung der Wissenschaftslehre' in *Fichtes Werke* (= repr. of *Johann Gottlieb Fichtes sämmtliche Werke*, ed. Immanuel Hermann Fichte, 8 vols, Berlin 1845–46 and *Johann Gottlieb Fichtes nachgelassene Werke*, ed. Immanuel Hermann Fichte, 3 vols, Bonn 1834–35), vol. I, Berlin 1971, pp. 519–34.

Fichte, Johann Gottlieb, 'Bericht über den Begriff der Wissenschaftslehre und die bisherigen Schicksale derselben' *in* Fichtes Werke (= repr. of *Johann Gottlieb Fichtes sämmtliche Werke*, ed. Immanuel Hermann Fichte, 8 vols, Berlin 1845–46 and *Johann Gottlieb Fichtes nachgelassene Werke*, ed. Immanuel Hermann Fichte, 3 vols, Bonn 1834–35), vol. VIII, Berlin 1971, pp. 361–407.

Fichte, Johann Gottlieb, 'Die Wissenschaftslehre' (1812) in *Fichtes Werke* (= repr. of *Johann Gottlieb Fichtes sämmtliche Werke*, ed. Immanuel Hermann Fichte, 8 vols, Berlin 1845–46 and *Johann Gottlieb Fichtes nachgelassene Werke*, ed. Immanuel Hermann Fichte, 3 vols, Bonn 1834–35), vol. X, Berlin 1971, pp. 315–492.

Finlay, Marike, *The Romantic Irony of Semiotics: Friedrich Schlegel and the Crisis of Representation*, Berlin 1988.

Fiorato, Pierfrancesco, '*Problemgeschichte* messianica: Nota sul metodo e il contenuto latente della dissertazione benjaminiana' in *Walter Benjamin: tra critica romantica e critica del Romanticismo*, eds Barnaba Maj, Davide Messina, Florence 2000, pp. 69–85.

Fiorato, Pierfrancesco, 'Die Erfahrung, das Unbedingte und die Religion: Walter Benjamin als Leser von *Kants Theorie der Erfahrung*' in *Hermann Cohen's Philosophy of Religion*, eds Stéphane Moses, Hartwig Wiedebach, Hildesheim 1992, pp. 71–84.

Flasch, Kurt, *Die geistige Mobilmachung: Die deutschen Intellektuellen und der Erste Weltkrieg*, Berlin 2000.

Frank, Manfred, *Einführung in die frühromantische Ästhetik: Vorlesungen*, Frankfurt/M. 1989.

Fuld, Werner, *Walter Benjamin: Zwischen den Stühlen: Eine Biographie*, Munich 1979.

Gangl, Manfred, Gérard Raulet (eds), *Intellektuellendiskurse in der Weimarer Republik: Zur politischen Kultur einer Gemengelage*, Darmstadt 1994.

Garber, Klaus, *Rezeption und Rettung: Drei Studien zu Walter Benjamin*, Tübingen 1987.

Garber, Klaus, 'Benjamin und das Barock: Ein Trauerspiel ohne Ende', *Euphorion*, 84 (1990), 207–12.

Geldsetzer, L. 'Problemgeschichte I' in *Historisches Wörterbuch der Philosophie*, eds Joachim Ritter, Karlfried Gründer, vol. 7, Darmstadt 1989, cols 1410–14.

Geulen, Eva, 'Zeit zur Darstellung: Walter Benjamins *Das Kunstwerk im Zeitalter seiner technischen Reduzierbarkeit*', *MLN*, 107:3 (1992), pp. 580–605.

Geuss, Raymond, *The Idea of a Critical Theory: Habermas and the Frankfurt School*, Cambridge 1981.

Goldstein, Moritz, 'German Jewry's Dilemma before 1914: The Story of a Provocative Essay' in *Leo Baeck Institute Yearbook* II, London 1957, pp. 236–54.

Greiffenhagen, Martin, *Das Dilemma des Konservatismus in Deutschland*, Munich 1971.

Grosse-Brockhoff, Annelen, *Das Konzept des Klassischen bei Friedrich und August Wilhelm Schlegel*, Cologne 1981.

Gundolf, Friedrich, *Romantiker*, Berlin-Wilmersdorf 1930.

Gundolf, Friedrich, *Shakespeare und der deutsche Geist*, Berlin 1911.

Güntert, Hermann, *Von der Sprache der Götter und Geister: Bedeutungsgeschichtliche Untersuchungen zur homerischen und eddischen Göttersprache*, Halle (Saale) 1921.

Habermas, Jürgen, *Technik und Wissenschaft als 'Ideologie'*, Frankfurt/M.1968; trans. 'Technology and Science as "Ideology"', Jeremy J. Shapiro, in Habermas,

Toward a Rational Society: Student Protest, Science, and Politics, London 1971, pp. 81–122.

Habermas, Jürgen, 'Bewußtmachende oder rettende Kritik – die Aktualität Walter Benjamins' in *Zur Aktualität Walter Benjamins*, ed. Siegfried Unseld, Frankfurt/M. 1972, pp. 174–223; trans. 'Consciousness-Raising or Redemptive Criticism: The Contemporaneity of Walter Benjamin', Philip Brewster, Carl H. Buchner, *New German Critique*, 17 (1979), pp. 30–59.

Hamacher, Werner, 'Afformative, Strike: Benjamin's "Critique of Violence"', trans. Dana Hollander, in *Walter Benjamin's Philosophy: Destruction and Experience*, eds Andrew Benjamin, Peter Osborne, London 1994, pp. 110–38.

Hamann, Johann Georg, *Aesthetica in nuce* in Hamann, *Sämtliche Werke*, ed. Josef Nadler, vol. 2, Vienna 1950, pp. 195–217.

Hamann, Johann Georg, 'Philologische Einfälle und Zweifel über eine akademische Preisschrift' in Hamann, *Sämtliche Werke*, ed. Josef Nadler, vol. 3, Vienna 1951, pp. 35–53.

Hamann, Johann Georg, 'Des Ritters von Rosencreuz letzte Willensmeynung' in Hamann, *Sämtliche Werke*, ed. Josef Nadler, vol. 3, Vienna 1951, pp. 25–33.

Hamann, Johann Georg, *Briefwechsel*, eds Arthur Henkel, Walther Ziesemer, 6 vols, Wiesbaden 1955–75.

Handelman, Susan A., *fragments of Redemption: Jewish Thought and Literary Theory*, Bloomington 1991.

Hansen, Miriam, 'Benjamin and Cinema: '"The Blue Flower in the Land of Technology"', *New German Critique*, 40 (1987), 179–224.

Haubold, S.,'Problemgeschichte III' in *Historisches Wörterbuch der Philosophie*, eds Joachim Ritter, Karlfried Gründer, vol. 7, Darmstadt 1989, cols 1416–17.

Hegel, Georg Wilhelm Friedrich, *Wissenschaft der Logik* in Hegel, *Gesammelte Werke*, vols 11 and 12, eds Friedrich Hogemann, Walter Jaeschke, Hamburg 1978 and 1981; trans. *Hegel's Science of Logic*, W. H. Johnston, L. G. Struthers, 2 vols, 3rd edition, London 1961.

Hegel, Georg Wilhelm Friedrich, *Phänomenologie des Geistes* in Hegel, *Gesammelte Werke*, vol. 9, eds Wolfgang Bonsiepen, Reinhard Heede, Hamburg 1980; trans. *Hegel's 'Phenomenology of Spirit'*, A. V. Miller, Oxford 1977.

Hegel, Georg Wilhelm Friedrich, *Vorlesungen über die Aesthetik*, vol. II, (= repr. of Hegel, *Sämtliche Werke: Jubiläumsausgabe in zwanzig Bänden*, vol. XIII) ed. Hermann Glockner, Stuttgart-Bad Cannstatt 1964; trans. *Aesthetics: Lectures on Fine Art*, vol. II, T. M. Knox, Oxford 1975.

Heidegger, Martin, Die *Kategorien- und Bedeutungslehre des Duns Scotus* in Heidegger, *Gesamtausgabe*, ed. Friedrich-Wilhelm von Herrmann, vol. 1, Frankfurt/M. 1978, pp. 189–411.

Heidegger, Martin, *Sein und Zeit*, 10th edition, Tübingen 1963; trans. *Being and Time*, John Macquarrie, Edward Robinson, Oxford 1962.

Heidegger, Martin, *Unterwegs zur Sprache*, 6th edition, Pfullingen 1979.

Heidegger, Martin, *Poetry, Language, Thought*, trans. Albert Hofstadter, New York 1971

Heil, Susanne, *'Gefährliche Beziehungen': Walter Benjamin und Carl Schmitt*, Stuttgart 1996.

Held, Hans Ludwig, 'Über Golem und Schem: Eine Studie aus der hebräischen Mystik', 2 parts, *Das Reich*, 1 (1916/17), 334–79 and 515–59.

Hellingrath, Norbert von, *Pindarübertragungen von Hölderlin: Prolegomena zu einer Erstausgabe*, Jena 1911.

Hillach, Ansgar, '"Ein neu entdecktes Lebensgesetz der Jugend": Wynekens Führergeist im Denken des jungen Benjamin' in *global benjamin: Internationaler Walter-Benjamin-Kongreß 1992*, eds Klaus Garber, Ludger Rehm, Munich 1999, vol. 2, pp. 872–90.

Hofmannsthal, Hugo von, *Das Schrifttum als geistiger Raum der Nation* in Hofmannsthal, *Gesammelte Werke, Reden und Aufsätze III (1925–1929)*, eds Bernd Schoeller, Ingeborg Beyer-Ahlert, Frankfurt/M. 1980, pp. 24–41.

Hölderlin, Friedrich, [*Neun Pindar-Kommentare*] in Hölderlin, *Sämtliche Werke: Kritische Textausgabe*, vol. 15, ed. D. E. Sattler, Darmstadt 1988, pp. 274–300.

Hölderlin, Friedrich, 'Über Religion' in Hölderlin, *Sämtliche Werke*, ed. Friedrich Beissner, vol. IV, Stuttgart 1961, pp. 275–81.

Hölderlin, Friedrich, *Gedichte nach 1800* (= Hölderlin, *Sämtliche Werke*, vol. II), ed. Friedrich Beissner, Stuttgart 1951.

Hölderlin, Friedrich, *Poems and Fragments: Bi-Lingual Edition with a Preface, Introduction and Notes*, trans. Michael Hamburger, London 1966.

Honneth, Axel, 'A Communicative Disclosure of the Past: On the Relation between Anthropology and Philosophy of History in Walter Benjamin' in *The Actuality of Walter Benjamin* (= *new formations*, 20, 1993), eds Laura Marcus, Lynda Nead, pp. 83–94.

Hübener, W., 'Problemgeschichte II' in *Historisches Wörterbuch der Philosophie*, eds Joachim Ritter, Karlfried Gründer, vol. 7, Darmstadt 1989, cols 1414–16.

Huge, Eberhard, *Poesie und Reflektion in der Ästhetik des frühen Friedrich Schlegel*, Stuttgart 1971.

Husserl, Edmund, *Philosophie als strenge Wissenschaft* in Husserl, *Gesammelte Werke*, vol. XXV, eds Thomas Nenon, Hans Rainer Sepp, Dodrecht 1987, pp. 3–62; trans. *Philosophy as Rigorous Science* in Husserl, *Phenomenology and the Crisis of Philosophy*, ed. and trans. Quentin Laver, New York 1965, pp. 71–147.

Husserl, Edmund, *Logische Untersuchungen*, vol. II, 5th edition, Tübingen 1968; trans. *Logical Investigations*, vol. II, J. N. Finlay, London 1970.

Husserl, Edmund, *Ideen zu einer reinen Phänomenologie und phänomenologischen Philosophie*, 3 vols (= *Husserliana*, vols III-V, eds Walter and Marly Biemel), The Hague 1950–52; trans. *Ideas: General Introduction to Pure Phenomenology*, W. R. Boyce Gibson, 4th edition, London 1967.

James, William, 'The Will to Believe' in James, *The Will To Believe and Other Essays in Popular Philosophy* and *Human Immortality: both books bound as one*, New York 1956, pp. 1–31.

Jauß, Hans Robert, *Kleine Apologie der ästhetischen Erfahrung: Mit kunstgeschichtlichen Bemerkungen von Max Imdahl*, Konstanz 1972.

Jauß, Hans Robert, *Ästhetische Erfahrung und literarische Hermeneutik*, Frankfurt/M. 1982.

Jennings, Michael W., *Dialectical Images: Walter Benjamin's Theory of Literary Criticism*, Ithaca 1987.

Kafka, Franz, 'Ein Brudermord' in Kafka, *Sämtliche Erzählungen*, ed. Paul Raabe, Frankfurt/M. 1972, pp. 161–4; trans. 'A Fratricide', Willa and Edwin Muir, in Kafka, *Collected Stories*, ed. Gabriel Josipovici, London 1993, pp. 190–2.

Kafka, Franz, 'Konvolut 1920' in Kafka, *Nachgelassene Schriften und Fragmente in der Fassung der Handschriften*, vol. II, ed. Jost Schillemeit, Frankfurt/M. 1992, pp.

223–362; 'Fragments from Note-Books and Loose Pages' in Kafka, *Wedding Preparations in the Country and Other Posthumous Prose Writings*, Ernst Kaiser and Eithne Wilkins, London 1954, pp. 218–413.

Kant, Immanuel, *Kritik der Urteilskraft*, ed. Karl Vorländer, Hamburg 1963; trans. *Critique of Judgment: Including The First Introduction*, Werner S. Pluhar, Indianapolis 1987.

Kambas, Chryssoula, *Walter Benjamin im Exil: Zum Verhältnis von Literaturpolitik und Ästhetik*, Tübingen 1983.

Kambas, Chryssoula, '*Politische Aktualität*: Walter Benjamin's Concept of History and the Failure of the French Popular Front', *New German Critique* 39 (1986), 87–98.

Kather, Regine, *'Über Sprache überhaupt und über die Sprache des Menschen': Die Sprachphilosophie Walter Benjamins*, Frankfurt/M. 1989.

Kierkegaard, Søren, *Either/Or*, trans. Walter Lowrie, London 1946.

Kierkegaard, Søren, *Fear and Trembling and The Sickness unto Death*, trans. Walter Lowrie, Princeton 1970.

Kim, Young-Ok, *Selbstporträt im Text des Anderen: Walter Benjamin's Kafka-Lektüre*, Frankfurt/M. 1995.

Kittsteiner, H. D., 'Walter Benjamins Historicism', trans. Jonathan Monroc, Irving Wohlfahrt, *New German Critique*, 39 (1986), 179–215.

Kittsteiner, Heinz-Dieter, 'The Allegory of the Philosophy of History in the Nineteenth Century', trans. Susan Gillespie with Daniela Franke, in *Walter Benjamin and the Demands of History*, ed. Micheal P. Steinberg , Ithaca 1996, pp. 41–61.

Koch, Gertrud, 'Cosmos in Film: On the Concept of Space in Walter Benjamin's "Work of Art" Essay', trans. Nancy Nenno, in *Walter Benjamin's Philosophy: Destruction and Experience*, eds Andrew Benjamin, Peter Osborne, London 1994, pp. 205–15.

Kohlenbach, Margarete, 'Vereinigung und Paradox: Anthropologische Bestimmungen in der Geschichtsphilosophie und Poetik des frühen Friedrich Schlegel' in *Literatur und Anthropologie um 1800*, eds Eda Sagarra, Jürgen Barkhoff, Munich 1991, pp. 141–60.

Kohlenbach, Margarete, 'The Desire for Objectivity: Walter Benjamin and Early Romanticism', *German Life and Letters*, New Series, vol. 48:1 (1995), 25–38.

Kohlenbach, Margarete, 'Heinles Verklärung: Zum esoterischen Subjektivismus in Walter Benjamins "Zwei Gedichte von Friedrich Hölderlin"' in *Wechsel der Orte: Studien zum Wandel des literarischen Geschichtsbewußtseins*, eds Irmela von der Lühe, Anita Runge, Göttingen 1997, pp. 138–57.

Königs, Liesel, 'Aktualitätstheorie' in *Historisches Wörterbuch der Philosophie*, eds Joachum Ritter, Karlfried Gründer, vol. 1, Darmstadt 1971, col. 145.

Königs, Liesel, 'Aktualismus' in *Historisches Wörterbuch der Philosophie*, eds Joachim Ritter, Karlfried Gründer, vol. 1, Darmstadt 1971, cols 144–5.

Korsch, Karl, *Marxismus und Philosophie*, Frankfurt/M. 1966.

Kosman, Aryeh, 'Acting: *Drama* as *Mimesis* of *Praxis*' in *Essays on Aristotle's Poetics*, ed. Amélie Oksenberg Rorty, Princeton 1992, pp. 51–72.

Kulenkampff, Jens, *Kants Logik des ästhetischen Urteils*, Frankfurt/M. 1978.

Kulenkampff, Jens, 'Ästhetische Erfahrung – oder was von ihr zu halten ist' in *Der Begriff der Erfahrung in der Philosophie des 20. Jahrhunderts*, eds Jürg Freudiger, Andreas Graeser, Klaus Petrus, Munich 1996, pp. 178–98.

Kupffer, Heinrich, *Gustav Wyneken: 1875–1964*, Stuttgart 1970.

Kutsch, E., 'Gottesurteil: II. In Israel' in *Die Religion in Geschichte und Gegenwart*, ed. Kurt Galling, 3rd edition, Tübingen 1986, vol. 2, cols 1808–9.

Lacoue-Labarthe, Philippe, Jean-Luc Nancy, *The Literary Absolute: The Theory of Literature in German Romanticism*, trans. Philip Barnard, Cheryl Lester, Albany 1988.

Lacoue-Labarthe, Philippe, 'Introduction to Walter Benjamin's "The Concept of Art Criticism in the German Romantics"', trans. David Ferris, *Studies in Romanticism*, 31 (1992), 421–32.

Lask, Emil, *Die Logik der Philosophie und die Kategorienlehre* in Lask, *Gesammelte Schriften*, ed. Eugen Herrigel, vol. 2, Tübingen 1923.

Lea, Kenneth, 'Traducing History: Benjamin, Language, Politics' in *Postmodernism and the Re-reading of Modernity*, eds Francis Barker, Peter Hulme, Margaret Iversen, Manchester 1992.

Leslie, Esther, *Walter Benjamin: Overpowering Conformism*, London 2000.

Lindner, Burkhardt, 'Engel und Zwerg: Benjamins geschichtsphilosophische Rätselfiguren und die Herausforderung des Mythos' in *'Was nie geschrieben wurde, lesen': Frankfurter Benjamin-Vorträge*, eds Lorenz Jäger, Thomas Regehly, Bielefeld 1992, pp. 236–65.

Lindner, Burkhardt, 'Benjamins Aurakonzeption: Anthropologie und Technik, Bild und Text' in *Walter Benjamin 1892–1940: zum 100. Geburtstag*, ed. Uwe Steiner, Bern 1992, pp. 217–48.

Lindner, Burkhardt, 'Derrida, Benjamin, Holocaust' in *Global Benjamin: Internationaler Walter-Benjamin-Kongreß 1992*, eds Klaus Garber, Ludger Rehm, Munich 1999, vol. 3, pp. 1691–723.

Lukács, Georg, *Die Theorie des Romans: Ein geschichtsphilosophischer Versuch über die Formen der großen Epik*, Neuwied 1971; trans. *The Theory of the Novel: A historico-philosophical essay on the forms of great epic literature*, Anna Bostock, London 1971.

Lukács, Georg, 'Tendenz oder Parteilichkeit', *Linkskurve*, 4:6 (1932), 13–21.

Lukács, Georg, 'Größe und Verfall des Expressionismus' in Lukács, *Essays über den Realismus* (= *Probleme des Realismus*, vol. I), Neuwied 1971, pp. 109–49.

Luther, Martin, 'Sermon und Eingang in das erste Buch Mose' in Luther, *Werke*, vol. 12, Graz 1966 (= repr. of the edition: Weimar 1891), pp. 435–52.

Luther, Martin, *Auslegung des ersten und zweiten Kapitels Johannis* in Luther, *Werke*, vol. 46, Graz 1967 (= repr. of the edition: Weimar 1912), pp. 538–789.

Maimonides, Moses, *The Guide for the Perplexed*, trans. M. Friedländer, 2nd edition, London 1956.

Makropoulos, Michael, 'Haltlose Souveränität: Benjamin, Schmitt und die Klassische Moderne in Deutschland' in *Intellektuellendiskurse in der Weimarer Republik*, eds. Manfred Gangl, Gérard Raulet, Darmstadt 1994, pp. 197–211.

Marx, Karl, 'Zur Kritik der Nationalökonomie – Ökonomisch-philosophische Manuskripte' in Marx, *Frühe Schriften*, vol. 1, eds Hans-Joachim Lieber, Peter Furth, Darmstadt 1971, pp. 506–665; trans. Martin Milligan, in *The Marx-Engels Reader*, ed. Robert C. Tucker, 2nd edition, New York 1978, pp. 66–125.

Mayer, Hans, 'Walter Benjamin und Franz Kafka: Bericht über eine Konstellation', *Literatur und Kritik*, 14 (1979), 579–97.

McCole, John, *Walter Benjamin and the Antinomies of Tradition*, Ithaca 1993.

McLaughlin, Brian P., Amélie Oksenberg Rorty (eds), *Perspectives on Self-Deception*, Berkeley 1988.

Meltzer, Françoise, 'Acedia and Melancholia' in *Walter Benjamin and the Demands of History*, ed. Michael P. Steinberg, Ithaca 1996, pp. 141–63.

Mendelssohn, Moses, *Jerusalem oder über religiöse Macht und Judentum* in Mendelssohn, *Schriften über Religion und Aufklärung*, ed. Martina Thom, Darmstadt 1989, pp. 351–458.

Menke, Bettine, *Sprachfiguren: Name-Allegorie-Bild nach Walter Benjamin*, Munich 1991.

Mennemeier, Franz Norbert, *Friedrich Schlegels Poesiebegriff dargestellt anhand der literaturkritischen Schriften: Die romantische Konzeption einer objektiven Poesie*, Munich 1971.

Menninghaus, Winfried, *Walter Benjamins Theorie der Sprachmagie*, Frankfurt/M. 1980.

Menninghaus, Winfried, *Unendliche Verdopplung: Die frühromantische Grundlegung der Kunsttheorie im Begriff absoluter Selbstreflexion*, Frankfurt/M. 1987.

Michel, Willy, *Ästhetische Hermeneutik und frühromantische Kritik: Friedrich Schlegels fragmentarische Entwürfe, Rezensionen, Charakteristiken und Kritiken (1795–1801)*, Göttingen 1982.

Micraelius, Johannes, 'communicatio verbalis' and 'communicatio realis' in Micraelius, *Lexicon Philosophicum Terminorum Philosophis Usitatorum*, Düsseldorf 1966 (= repr. of the 2nd edition: Stettin 1662), cols 302–3.

Mieth, Dietmar, 'Die Fremdheit Gottes in Mystik und Moderne' in *Religion zwischen Gewalt und Beliebigkeit*, eds Dietmar Mieth, Hadewych Snijdewind, Tübingen 2001, pp. 107–22.

Moretti, Franco, *Signs Taken for Wonder: Essays in the Sociology of Literary Forms*, trans. Susan Fischer, David Forgacs, David Miller, revised edition, London 1988.

Moses, Stéphane, 'Walter Benjamin and Franz Rosenzweig' in *Benjamin: Philosophy, Aesthetics, History*, ed. Gary Smith, Chicago 1989, pp. 228–46.

Moses, Stéphane, 'Brecht und Benjamin als Kafka-Interpreten' in *Juden in der deutschen Literatur: Ein deutsch-israelisches Symposion*, eds Stéphane Moses, Albrecht Schöne, Frankfurt/M. 1986, pp. 237–56.

Mueller, Roswitha, 'Learning for a new society: the *Lehrstück*' in *Brecht*, eds Peter Thomson, Glendyr Sacks, Cambridge 1994, pp. 79–95.

Müller, Harro, 'Einige Anmerkungen zu Benjamins Historismuskritik' in *global benjamin: Internationaler Walter-Benjamin-Kongreß 1992*, eds Klaus Garber, Ludger Rehm, Munich 1999, vol. 1, pp. 152–60.

Nägele, Rainer, *Theater, Theory, Speculation: Walter Benjamin and the Scenes of Modernity*, Baltimore 1991.

Nancy, Jean-Luc, Philippe Lacoue-Labarthe, *The Literary Absolute: The Theory of Literature in German Romanticism*, trans. Philip Barnard, Cheryl Lester, Albany 1988.

Natorp, Paul, 'Kant und die Marbuger Schule', *Kant-Studien*, 17 (1912), 193–221.

Natorp, Paul, *Einleitung in die Psychologie nach kritischer Methode*, Freiburg i. B. 1888.

Neuhouser, Frederick, *Fichte's Theory of Subjectivity*, Cambridge 1990.

Nietzsche, Friedrich, *Die Geburt der Tragödie aus dem Geiste der Musik* in Nietzsche, *Werke in drei Bänden*, ed. Karl Schlechta, vol. 1, Munich 1966; trans. *The Birth of*

Tragedy or 'Hellenism' and 'Pessimism', W. M. A. Haussmann, in Nietzsche, *The Complete Works*, ed. O. Levy, vol. 3, Edinburgh 1909.

Nottarp, H., 'Gottesurteil: I Rechtsgeschichtlich' in *Die Religion in Geschichte und Gegenwart*, ed. Kurt Galling, 3rd edition, Tübingen 1986, vol. 2, cols 1807–8.

Novalis, [Fichte-Studien] in Novalis, *Schriften*, ed. Richard Samuel, vol. 2, Stuttgart 1960, pp. 29–296.

Novalis, 'Logologische Fragmente' in Novalis, *Schriften*, ed. Richard Samuel, vol. 2, Stuttgart 1960, pp. 522–63.

Novalis, 'Vermischte Bemerkungen' in Novalis, *Schriften*, ed. Richard Samuel, vol. 2, Stuttgart 1960, pp. 412–70.

Oetinger, Friedrich, Christoph, *Die Lehrtafel der Prinzessin Antonia*, eds Reinhard Breymayer, Friedrich Häussermann, Berlin 1977.

Pascal, Blaise, *Pensées* in Pascal, *Ovres Complètes*, ed. Louis Lafuma, Paris 1963, pp. 493–641; trans. *Pensées: Notes on Religion and Other Subjects*, John Warrington, London 1973.

Pears, David, *Motivated Irrationality*, Oxford 1984.

Pfotenhauer, Helmut, '"Eine Puppe in türkischer Tracht": Zur Verbindung von historischem Materialismus und Theologie beim späten Benjamin' in *Materialien zu Benjamins Thesen 'Über den Begriff der Geschichte'*, ed. Peter Bulthaupt, Frankfurt/M. 1975, pp. 254–91.

Pfotenhauer, Helmut, 'Benjamin und Nietzsche' in *Walter Benjamin im Kontext*, ed. Burkhardt Lindner, 2nd edition, Königstein 1985, pp.100–26.

Pfotenhauer, Helmut, *Ästhetische Erfahrung und gesellschaftliches System: Untersuchungen zu Methodenproblemen einer materialistischen Literaturanalyse am Spätwerk Walter Benjamins*, Stuttgart 1975.

Pivecka, Alexander, *Die künstliche Natur: Walter Benjamin's Begriff der Technik*, Frankfurt/M. 1993.

Polheim, Karl Konrad, *Die Arabeske: Ansichten und Ideen aus Friedrich Schlegels Poetik*, Munich, 1966.

Prickett, Stephen, *Words and 'The Word': Language, poetics and biblical interpretation*, Cambridge 1988.

Pulver, Max, *Romantische Ironie und Romantische Komödie*, St. Gallen 1912 (PhD diss. Freiburg i. Br.).

Rabinbach, Anson, 'Between Enlightenment and Apocalypse: Benjamin, Bloch and Modern German Jewish Messianism', *New German Critique*, 34 (1985), 78–124.

Raulet, Gérard, Manfred Gangl (eds), *Intellektuellendiskurse in der Weimarer Republik: Zur politischen Kultur einer Gemengelage*, Darmstadt 1994.

Recki, Birgit, *Aura und Autonomie: Zur Subjektivität der Kunst bei Walter Benjamin und Theodor W. Adorno*, Würzburg 1988.

Riegl, Alois, *Spätrömische Kunstindustrie*, 2nd edition, Vienna 1927.

Reijen, Willem van, Norbert Bolz, *Walter Benjamin*, Frankfurt/M. 1991.

Ringer, Fritz K., *The Decline of the German Mandarins: The German Academic Community, 1890–1933*, Cambridge (Mass.) 1969.

Rochlitz, Rainer, *The Disenchantment of Art: The Philosophy of Walter Benjamin*, New York 1996.

Rorty, Amélie Oksenberg, Brian P. McLaughlin, (eds), *Perspectives on Self-Deception*, Berkeley 1988.

Rose, Gillian, 'Hermann Cohen – Kant among the Prophets' in Rose, *Judaism and Modernity*, Oxford 1993, pp. 111–25.

Rose, Gillian, 'Walter Benjamin – Out of the Sources of Modern Judaism', in Rose, *Judaism and Modernity*, Oxford 1993, pp. 175–210.

Rosenzweig, Franz, *Das älteste Systemprogramm des deutschen Idealismus: Ein handschriftlicher Fund*, Heidelberg 1917.

Rosenzweig, Franz, *Der Stern der Erlösung*, Frankfurt/M. 1993; trans. *The Star of Redemption*, William Hallo, London 1971.

Rothe, Richard, *Zur Dogmatik*, Gotha 1863.

Rüffer, Ulrich, 'Taktilität und Nähe' in *Antike und Moderne: Zu Walter Benjamins 'Passagen'*, eds Norbert W. Bolz, Richard Faber, Würzburg 1986, pp. 181–90.

Rumpf, Michael, *Aporien und Apologie: Zur Kritik Walter Benjamins und seiner Rezeption*, Cuxhaven 1997.

Russell, Bertrand, Alfred North Whitehead, *Principia Mathematica*, 2nd edition, 3 vols, Cambridge 1960, 1959, 1927.

Santner, Eric L., *On the Psychotheology of Everyday Life: Reflections on Freud and Rosenzweig*, Chicago 2001.

Schelling, Friedrich Wilhelm Josef, *Philosophische Untersuchung über das Wesen der menschlichen Freiheit und die damit zusammenhängenden Gegenstände* in Schelling, *Schriften von 1806–1813* (= *Ausgewählte Werke*, vol. 9), Darmstadt 1968, pp. 275–360; trans. *Schelling: Of Human Freedom*, ed. and trans. James Gutmann, Chicago 1936.

Schelling, Friedrich Wilhelm Josef, *System des transcendentalen Idealismus* in Schelling, *Schriften von 1799–1801* (= *Ausgewählte Werke*, vol. 7), Darmstadt 1967, pp. 327–634; trans. *System of Transcendental Idealism (1800)*, Peter Heath, Charlottesville 1978.

Schelling, Friedrich, Wilhelm, Josef von, *Ideen zu einer Philosophie der Natur* in Schelling, *Sämtliche Werke*, ed. K. F. A. Schelling, vol. II, Stuttgart 1857, pp. 1–343.

Schiller, Friedrich, *On the Aesthetic Education of Man in a Series of Letters*, ed. and trans. Elizabeth M. Wilkinson, L. A. Willoughby, Oxford 1967.

Schings, Hans-Jürgen, 'Consolatio Tragoediae: Zur Theorie des barocken Trauerspiels' in *Deutsche Dramentheorie: Beiträge zu einer historischen Poetik des Dramas in Deutschland*, ed. Reinhold Grimm vol. 1, Frankfurt/M. 1971, pp. 1–44.

Schings, Hans-Jürgen, 'Walter Benjamin, das barocke Trauerspiel und die Barockforschung' in *Daß eine Nation die ander verstehen möge*, eds Norbert Honsza, Hanz-Gert Roloff, Amsterdam 1988, pp. 663–76.

Schleiermacher, Friedrich, *Der christliche Glaube, nach den Grundsätzen der evangelischen Kirche im Zusammenhange dargestellt* in Schleiermacher, *Kritische Gesamtausgabe*, ed. Hermann Peiter, vols I.7,1 and I.7,2, Berlin 1980.

Schlüter, D., 'Akt/Potenz' in *Historisches Wörterbuch der Philosophie*, eds Joachim Ritter, Karlfried Gründer, vol. 1, Darmstadt 1971, cols 134–42.

Schmidt, Jochen, *Die Geschichte des Genie-Gedankens 1750–1945*, 2 vols, Darmstadt 1985.

Schnädelbach, Herbert, *Philosophy in Germany: 1831–1933*, Cambridge 1984.

Scholem, Gershom, *The Messianic Idea in Judaism and Other Essays on Jewish Spirituality*, London 1971.

Scholem, Gershom, 'The Name of God and the Linguistic Theory of the Kabbala', 2 parts, *Diogenes* 79 (1972), 59–80, and 80 (1972), 165–94.

Scholem, Gershom, *Zur Kabbala und ihrer Symbolik*, Frankfurt/M. 1973.

Scholem, Gershom, *Walter Benjamin: Geschichte einer Freundschaft*, Frankfurt/M. 1975; trans. *Walter Benjamin: The Story of a Friendship*, Harry Zohn, Philadelphia 1981.

Schöttker, Detlev, *Konstruktiver Fragmentarismus: Form und Rezeption der Schriften Walter Benjamins*, Frankfurt/M. 1999.

Schwarz Wentzer, Thomas, *Bewahrung der Geschichte: Die hermeneutische Philosophie Walter Benjamins*, Bodenheim 1998.

Smith, Gary, '"Das Jüdische versteht sich von selbst": Walter Benjamins frühe Auseinandersetzung mit dem Judentum', *Deutsche Vierteljahrsschrift für Literaturwissenschaft und Geistesgeschichte*, 65 (1991), 318–34.

Sontheimer, Kurt, 'Der Tatkreis', *Vierteljahrshefte für Zeitgeschichte*, 7:3 (1959), 229–60.

Sontheimer, Kurt, *Antidemokratisches Denken in der Weimarer Republik: Die politischen Ideen des deutschen Nationalismus zwischen 1918 und 1933*, Munich 1962.

Speth, Rudolf, *Wahrheit und Ästhetik: Untersuchungen zum Frühwerk Walter Benjamins*, Würzburg 1991.

Steiner, Uwe, *Die Geburt der Kritik aus dem Geiste der Kunst: Untersuchungen zum Begriff der Kritik in den frühen Schriften Walter Benjamins*, Würzburg 1989.

Steiner, Uwe, '*Zarte Empirie*: Überlegungen zum Verhältnis von Urphänomen und Ursprung im Früh- und Spätwerk Walter Benjamins' in *Antike und Moderne: Zu Walter Benjamins 'Passagen'*, eds Norbert W. Bolz, Richard Faber, Würzburg 1986, pp. 20–40.

Steiner, Uwe, 'Allegorie und Allergie: Bemerkungen zur Diskussion um Benjamins Trauerspielbuch in der Barockforschung', *Daphne* 18 (1989), 641–701.

Stoessel, Marleen, *Aura: Das vergessene Menschliche: Zu Sprache und Erfahrung bei Walter Benjamin*, Munich 1983.

Strachey, Lytton, *Eminent Victorians*, London 1986.

Strohschneider-Kohrs, Ingrid, *Die romantische Ironie in Theorie und Gestaltung*, 2nd edition, Tübingen 1977.

Swinburne, Richard, *Revelation: From Metaphor to Analogy*, Oxford 1992.

Thornhill, Christopher J., *Walter Benjamin and Karl Kraus: Problems of a 'Wahlverwandtschaft'*, Stuttgart 1996.

Thornhill, Chris, *Political Theory in Modern Germany: An Introduction*, Cambridge 2000.

Trappe, Tobias, 'Selbstoffenbarung' in *Historisches Wörterbuch der Philosophie*, eds Joachim Ritter, Karlfried Gründer, vol. 9, Darmstadt 1995, cols 503–10.

Uehling, Theodore Edward, Jr., *The Notion of Form in Kant's 'Critique of Aesthetic Judgment'*, The Hague 1971.

Unger, Erich, *Politik und Metaphysik*, ed. Manfred Voigts, Würzburg 1989 (= repr. of the edition: Berlin 1921).

Voigts, Manfred, 'Walter Benjamin und Erich Unger: Eine jüdische Konstellation' in *global benjamin: Internationaler Walter-Benjamin-Kongreß 1992*, eds Klaus Garber, Ludger Rehm, Munich 1999, vol. 2, pp. 839–55.

Volkmann-Schluck, Karl Heinz, 'Novalis' magischer Idealismus' in *Die deutsche*

Romantik: Poetik, Formen und Motive, ed. Hans Steffen, 3rd edition, Göttingen 1978, pp. 45–53.

Vries, Josef de, *Grundbegriffe der Scholastk*, 2nd edition, Darmstadt 1983.

Weber, Max, 'Wissenschaft als Beruf' in Weber, *Gesammelte Aufsätze zur Wissenschaftslehre*, 5th edition, ed. Johannes Winckelmann, Tübingen 1982, pp. 582–613; trans. 'Science as a Vocation' in *From Max Weber: Essays in Sociology*, 2nd edition, eds and trans. H. H. Gerth, C. Wright Mills, London 1991, pp. 129–56.

Weidmann, Heiner, 'Geistesgegenwart: Das Spiel in Walter Benjamins Passagenarbeit', *MLN*, 107 (1992), 521–47.

Weigel, Sigrid, *Body- and Image-Space: Re-reading Walter Benjamin*, trans. Georgina Paul with Rachel McNicholl, Jeremy Gaines, London 1996.

Werbick, Jürgen, 'Das Medium ist die Botschaft: Über einige wenig beachtete Implikationen des Begriffs der "Selbstoffenbarung Gottes" – mit Blick auf die Auseinandersetzung um die fundamentalistische Versuchung im Christentum' in *Offenbarungsanspruch und fundamentalistische Versuchung*, ed. Werbick, Freiburg 1991, pp. 187–245.

Whitehead, Alfred North, Bertrand Russell, *Principia Mathematica*, 2nd edition, 3 vols, Cambridge 1960, 1959, 1927.

Wiesenthal, Lieselotte, *Zur Wissenschaftstheorie Walter Benjamins*, Frankfurt/M. 1973.

Willett, John (ed. and trans.), *Brecht on Theatre: The Development of an Aesthetic*, London 1986.

Windelband, Wilhelm, *Die Geschichte der neueren Philosophie in ihrem Zusammenhange mit der allgemeinen Kultur und den besonderen Wissenschaften*, 2 vols, 3rd edition, Leipzig 1904.

Wippel, John F., 'Metaphysics' in *Aquinas*, eds Norman Kretzmann, Eleonore Stump, Cambridge 1993, pp. 85–127.

Witte, Bernd, *Walter Benjamin: Der Intellektuelle als Kritiker*, Stuttgart 1976.

Witte, Bernd, *Walter Benjamin*, Reinbek 1985.

Witte, Bernd, '"Die Welt allseitiger und integraler Aktualität": Die Säkularisierung jüdischer Motive im Denken Walter Benjamins', *Der Deutschunterricht*, 37:3 (1985), 26–37.

Wohlfarth, Irving, '"Immer radikal, niemals konsequent ... ": Zur theologisch-politischen Standortsbestimmung Walter Benjamins' in *Antike und Moderne: Zu Walter Benjamins 'Passagen'*, eds Norbert W. Bolz, Richard Faber, Würzburg 1986, pp. 116–37.

Wohlfarth, Irving, 'Smashing the Kaleidoscope: Walter Benjamin's Critique of Cultural History' in *Walter Benjamin and the Demands of History*, ed. Michael P. Steinberg, Ithaca 1996, pp. 190–205.

Wolin, Richard, *Walter Benjamin: An Aesthetic of Redemption*, 2nd edition, Berkeley 1994.

Wolin, Richard, *Labyrinths: Explorations in the Critical History of Ideas*, Amherst 1995.

Wyneken, Gustav, *Schule und Jugendkultur*, 2nd edition, Jena 1914.

Wyneken, Gustav, 'Wandervogel und Freie Schulgemeinde' in Wyneken, *Der Kampf für die Jugend: Gesammelte Aufsätze*, Jena 1919, pp. 128–38.

Index

Bold type is used for references to the explanation or translation of non-English terms